Campanha Gaúcha:
A Brazilian Ranching
System, 1850–1920

Campanha Gaúcha

A BRAZILIAN RANCHING
SYSTEM, 1850–1920

Stephen Bell

STANFORD UNIVERSITY PRESS
STANFORD, CALIFORNIA
1998

Stanford University Press
Stanford, California
© 1998 by the Board of Trustees of the
Leland Stanford Junior University
Printed in the United States of America
CIP data appear at the end of the book

Published with the assistance of
the CRB Foundation

For farmer forebears, especially my late father

Acknowledgments

The geographer Carl Sauer argued that we can only learn and synthesize the particular character of foreign culture areas slowly, through prolonged fieldwork. My experience certainly reflects this lesson. Thus my first debt of thanks goes to the Social Sciences and Humanities Research Council of Canada, whose ongoing support made possible much of the South American research vital to this book.

I extend my thanks to all those who helped shape this work in their various ways, including my examiners and manuscript reviewers. The book began life as a doctoral thesis written for the Department of Geography at the University of Toronto. I am grateful to that university for the teaching I received there and for the fine collections of its Robarts Library. In particular, I thank Jock Galloway for the stimulus of his course on the cultural-historical geography of Latin America and for his careful supervision of my student work. Among the many teachers of geography I had earlier at Oxford University, my greatest debts are to Isobel Cosgrove, sometime lecturer in geography at The Queen's College, and to the late Frank Emery. The latter did more to kindle my interest in historical geography than he ever knew.

In work involving research in several countries, I have accumulated debts to many people. Any attempt to list those who assisted in the archives, specialized institutions, and libraries where I worked (in Brazil, Canada, England, and Uruguay) would surely fail by serious omission. All my requests to view material met with respectful attention, and I benefited greatly from that. In Rio Grande do Sul, I particularly enjoyed the enthusiasm of the staff in the Arquivo Histórico and the open welcome of the Instituto Histórico e Geográfico. Three people in Porto Alegre lent special assistance. The geographer Raphael Copstein helped to orient my early field research. Writer and translator Carmen Vera Cirne Lima helped me work on inventories in the Arquivo Público; in the process, she and her family became friends for life. Local historian

Carlos Reverbel granted advice from the peace of his private library, a collection rich in materials on Rio Grande do Sul.

A series of University of Toronto Open Fellowships financed much of my graduate work. Grants from the same university's Centre for International Studies and the International Development Research Council (administered by the Canadian Association of Latin American and Caribbean Studies) helped with the expensive matter of travel to and from Brazil.

The Department of Geography at McGill University has provided a welcome base for my research and I appreciate support received there, including the privilege of teaching some very fine students. As the project moved from thesis to book, the keen encouragement of John Wirth of the Department of History, Stanford University, helped me enormously. The CRB Foundation provided valuable, practical help with meeting the costs of publication. I thank the copyright holders for permission to draw material from my articles in *The Americas* and the *Journal of Historical Geography*. Eric Ross and Richard Bachand drew the maps, under my supervision. All translations and errors in this work are mine.

It has been a pleasurable experience working with the staff of Stanford University Press. I am especially grateful to Norris Pope, the director, for his guidance—always cheerfully given. I also wish to thank John Feneron and Lynn Stewart, my editor, for their careful work.

Finally, I commend Ann Dadson for constant encouragement. Without her support *em casa*, this book might never have surfaced to reach publication.   S.B.

Contents

Tables and Figures

Tables

Figures

Note on Currency, Units, and Spelling

Brazil's monetary unit during the period of this study was the *mil-réis*, written 1$000. Large sums were expressed in *contos*. A conto equaled 1,000 mil-réis (written 1:000$000). Sometimes I have given the approximate equivalent value of a sum in pounds sterling, the leading currency during the period of the study, so that readers may make ready comparisons with other parts of the world economy. I have used the data on sterling/mil-réis exchange rates presented in Oliver Onody, *A inflação brasileira, 1820–1958* (Rio de Janeiro, 1960), pp. 22–23.

Units in this study are usually given in metric terms, or regional measures with their metric equivalents. Where I have used non-metric units, such as acres when comparing the sizes of landed estates in Rio Grande do Sul with those of the British aristocracy, my rationale has been to help a specific cross-cultural comparison.

Portuguese has undergone many changes in orthography. I have modernized the names of geographical features and of entities (except for ranches) in the text. Titles of books and documents are cited using their original spelling. Quotations retain the original orthography. I have tended to retain the original spelling of names of persons. Exceptions to this come with the names of figures well known by modernized spelling in the literature (for example, Getúlio Dorneles Vargas, rather than Getúlio Dornelles Vargas). Achieving consistency with Portuguese spellings is a major challenge for Brazilianist and Brazilian alike; thus I plead the forgiveness of the reader where I have erred on this subject.

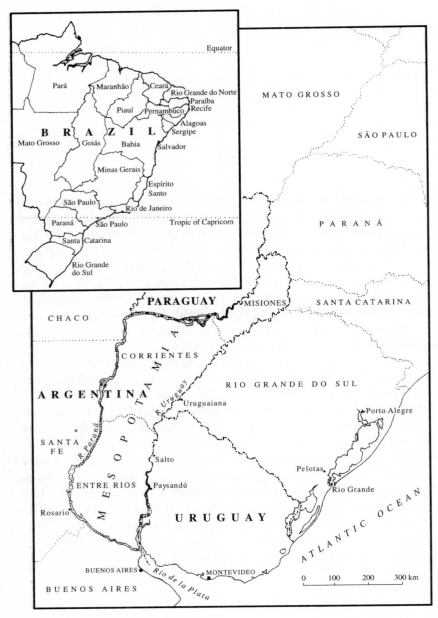

Location of Rio Grande do Sul.

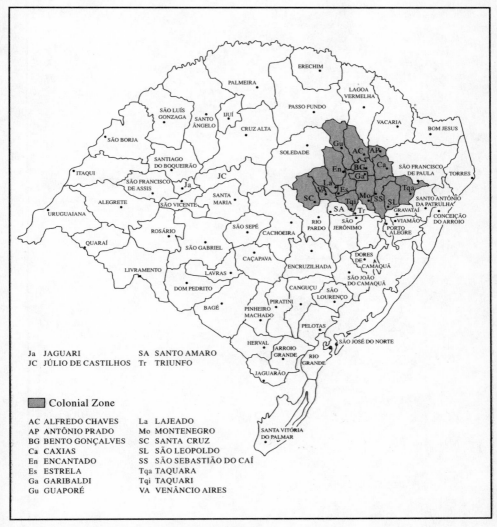

ERECHIM

PALMEIRA

LAGOA
VERMELHA

SÃO LUÍS
GONZAGA
SANTO
ÂNGELO

IJUÍ

PASSO FUNDO

VACARIA

BOM JESUS

SÃO BORJA

CRUZ ALTA

SOLEDADE

Gu

AC AP

SÃO FRANCISCO
DE PAULA

TORRES

SANTIAGO
DO BOQUEIRÃO

JC

BG Ca

En

SÃO FRANCISCO
DE ASSIS

Ja

Ga

La

Mo

Tqa

ITAQUI

SANTA
MARIA

VA

Es

SS

SL

SANTO ANTÔNIO
DA PATRULHA

ALEGRETE

SÃO VICENTE

SC

Tqi

GRAVATAÍ

URUGUAIANA

SÃO SEPÉ

SA

Tr

SÃO
JERÔNIMO

VIAMÃO

CONCEIÇÃO
DO ARRÓIO

ROSÁRIO

CACHOEIRA

RIO
PARDO

PORTO
ALEGRE

QUARAÍ

SÃO GABRIEL

CAÇAPAVA

ENCRUZILHADA

DORES
DE
CAMAQUÃ

LIVRAMENTO

LAVRAS

CANGUÇU

SÃO JOÃO
DO CAMAQUÃ

DOM PEDRITO

SÃO
LOURENÇO

BAGÉ

PINHEIRO
MACHADO

PIRATINI

PELOTAS

HERVAL

ARROIO
GRANDE

SÃO JOSÉ DO NORTE

JAGUARÃO

RIO
GRANDE

Ja JAGUARI   SA SANTO AMARO
JC JÚLIO DE CASTILHOS Tr TRIUNFO

SANTA VITÓRIA
DO PALMAR

Colonial Zone

AC ALFREDO CHAVES   La LAJEADO
AP ANTÔNIO PRADO   Mo MONTENEGRO
BG BENTO GONÇALVES   SC SANTA CRUZ
Ca CAXIAS   SL SÃO LEOPOLDO
En ENCANTADO   SS SÃO SEBASTIÃO DO CAÍ
Es ESTRELA   Tqa TAQUARA
Ga GARIBALDI   Tqi TAQUARI
Gu GUAPORÉ   VA VENÂNCIO AIRES

Counties (*Municípios*) in Rio Grande do Sul, 1920.

Campanha Gaúcha:
A Brazilian Ranching
System, 1850–1920

Introduction

RANCHING HAS MADE a vital, if increasingly controversial, contribution to Brazil's development, whether from Marajó Island in the Amazon basin to the semi-arid *sertão* of the northeast, or from its far west in the Mato Grosso to its extreme south in Rio Grande do Sul (see the Frontispiece).[1] Of the many pastoral systems within this country's vast territory, that of Rio Grande do Sul is regarded as the most important, a reflection of the economic power achieved on the country's only portion of temperate grasslands. Even though its economy in the twentieth century is diversified and increasingly industrial, outsiders still frequently regard the state as synonymous with ranching. Its other major contribution in the popular mind has been to the political architecture of modern Brazil; at the head of any roster of the major political figures the state has supplied stands Getúlio Vargas (1883–1954), scion of a prominent ranching family.[2]

Temperate Brazil is a different Brazil, as guidebooks never cease to stress. In some of its structural elements, it seems another country. At a superficial level, the ranching zones of Rio Grande do Sul, especially the Campanha, the region of rolling hills in the southwestern part of the state (see Figure 1.1 in Chapter 1), have shared a great deal in their historical geography, as well as their physical conditions, with the grasslands of Argentina and Uruguay to the south.[3] And during colonial times, Brazil's southern borderlands had a rural economy similarly based on the extensive herding of feral, or semi-wild, animals, hunted down by the nomadic cowboys known as *gaúchos*.[4] The breadth of common features in the social and economic development of the Spanish and Portuguese zones around the Río de la Plata estuary is impressive. These regions invite comparison.[5]

On the other hand, Rio Grande do Sul did not mirror the Río de la Plata in the patterns of change of its ranching economy. One obvious example is that Brazil was still using slave labor decades after Argentina and Uruguay had abolished the institution of slavery. Moreover, technical refinement in ranching, seen in such visible symbols of economic progress as the barbed-wire fence or the pedigree bull, came generally later in southernmost Brazil than within the Plata. In the light of these factors, scholars examining the economic transformation of pastoral activities in southern South America have viewed Rio Grande do Sul as a laggard region. In a synthesis of the regional geography of Brazil, for example, J. H. Galloway has noted the following: "To the south, in Uruguay and on the pampas of Argentina, there had been a revolution in livestock-ranching during the second half of the nineteenth century . . . but in Rio Grande, for reasons not yet clear, this revolution had scarcely begun by 1900."[6] The core of the investigation that follows is to clarify the character of Riograndense ranching modernization in the face of such characterization of retarded development in the region.

This book engages in the search for the mechanisms that explain landscape change on the grasslands of southern Brazil. There were marked regional differences in the timing and character of how the southern South American ranching economies were drawn into the international economy. The broad outline of John Friedmann's theoretical model of polarized development seems directly relevant to the historical experiences of southern South America.[7] First in importance in his model is the proposition that the diffusion of cultural and economic change emanates from centers to peripheries.[8] In general, the processes of modernization can be viewed as originating in the Río de la Plata and in São Paulo. Major cities like Montevideo, and regional centers such as Pelotas or Porto Alegre, acted as the relay centers for a more modern worldview. On account of the modernization of the last century, the contrasts between city and country became all the sharper. In the interior, the visible results of the Europeanizing liberal effort sometimes appeared only wafer-thin.[9] As ranching innovations confronted a variety of traditional situations, there was differential adoption of ranch technology. Cultural obstacles sometimes stood in the way, but also important were structural factors, barely visible in the modern historical literature on the region. Keeping a research focus on the historical-geographical circumstances of ranching innovation should help to clarify the important uneven geography of effective demand for widespread structural change across the many subregions of the southern South American ranching region.

Torn between two metropolitan regions, the Río de la Plata and south-central Brazil, Rio Grande do Sul is a particularly interesting region to consider within the core-periphery model of Latin American development. Its temperate grasslands, lying at a northern margin of the great pampa grassland system, were tied by their physical geography to the Plata; from this viewpoint, the Campanha was principally a periphery of Montevideo and not of Rio de Janeiro. At the same time, distance, infrastructure (notably the seasonally poor navigability of the Uruguay River), and culture isolated the region from the cities of the estuary. Politically, Rio Grande do Sul has always been part of Brazil, whose institutions have linked it with the rest of its country. During the nineteenth century, the Brazilian nation's oft-noted political triumph was that the Empire managed to hold the country together as a single entity. In economic terms, however, the sense of nation was a very loose one. In some ways imperial Brazil behaved more like a series of regional economies, yet the province of Rio Grande do Sul was important in relation to such interdependencies as were present. At the middle of the nineteenth century, its ports had the largest share by value of exports entering coastal traffic.[10] This trade pattern strengthens Fernando Henrique Cardoso's thesis that Rio Grande do Sul constituted a subsidiary economy, molded in large part by the exigencies of the coffee- and sugar-exporting zones of Brazil that consumed Riograndense salt-beef and agricultural products.[11]

In the period under study, Rio Grande do Sul was a marginal area in many ways. The region was marginal in a territorial sense within the vast space of Brazil. The subsidiary character of much of the pastoral economy that had developed there in past centuries also translated into marginality within the political life of the country. Although ranching had been an important activity in Brazil even in colonial times, the economic interests of the pastoral sector of Rio Grande do Sul never weighed anything like so heavily at the national level as ranching did in Argentina and Uruguay.[12] Throughout the period of this study, the politicians of the southern frontier held limited powers in national decision making. What significance did this political marginality hold for the evolution of the rural society and economy?

Like Albert Hirschman's entrepreneurs, researchers have tended to underestimate opportunities in the periphery.[13] A strong emphasis on dependency models in Latin American studies, at its peak in the 1970's, brought with it a concern for those regions most visibly molded by export sectors.[14] As the fashion has changed to include looking at societies from below as well as from above, research has started to examine the consequences of export-led development for regions not producing directly for foreign markets. Rio Grande do Sul exhibits characteristics of both of these frameworks for analyzing regions but sits comfortably in neither exclusive mold.

While the period 1850–1920 cuts across two major phases in Brazilian political history, the Empire (1822–89) and the Old Republic (1889–1930), there is still a coherence to it, stemming in large part from the distinctive world setting, which saw Northern Hemisphere technology arriving in the south on an entirely new scale.[15] These decades represented the peak of British power in Latin America. It was in the second half of the nineteenth century that ranching developed into what David Grigg has termed a "major agricultural system."[16] The technology of the Industrial Revolution allowed for the development of important new stock-raising regions, such as the Great Plains of the United States. In longer-established American ranching zones, such as the Campanha of Rio Grande do Sul, similar technology provided for the intensification of production and processing. The leading factor explaining the transformation of ranching techniques was the growing demand for imported food as well as for industrial raw materials, such as hides and wool, within the economies of the industrializing regions of Europe and the northeastern United States. While development during this period did not follow linear trends (it was disrupted, for example, by the American Civil War and by the Great Depression after 1873, as well as by countless regional interruptions to the workings of the international economy), this was the time when Britain reached the apogee of its power, and thus of its commercial influence in the world. By the middle of the nineteenth century, Britain, the world's principal economic power, could no longer feed itself and came to depend increasingly on imported food. This happened despite the impressive modernization of the domestic agricultural sector. By 1914, this leading economy of the world was facing intense competition in industrial production from other parts of Europe and from the United States. Given this competition for markets, the whole world could not be turned, in Eric Hobsbawm's phrasing, into a "kind of planetary system circling round the economic sun of Britain."[17]

Nevertheless, the growth of a vast international trade in foodstuffs after 1870 meant that the development of many regions of the world exhibited a form closely resembling what the liberal economists imagined as the ideal. Each section of what were later termed the "underdeveloped" regions of the globe supplied to the metropolitan countries those commodities which geography had best favored it to produce.

Technological changes were clearly of the highest importance in these international developments. Until the railway and steamship came onto the scene, the scope for international trade had been limited.[18] After 1875, world trade trebled in volume in the period before World War I. And the later decades of the nineteenth century saw investment in the agricultural development of the temperate areas on a vast scale. Again, Hobsbawm has summarized the trend: "The 'white' dominions (Canada, Australia, New Zealand, South Africa) raised their share [of British investment] from twelve per cent in the 1860's to almost thirty per cent in the 1880's; and if we include Argentina, Chile and Uruguay as 'honorary' dominions—their economies were not dissimilar—the rise in these outlets for capital export is even more striking."[19] In the developing, industrializing nations, Britain could not expect to maintain her economic supremacy, or monopoly, intact. But over a large part of the underdeveloped world, there was the legitimate hope that a permanent complementarity of economies might be either attained or maintained. This conviction was only seriously brought to question with the shock to the world economy occasioned by the events of 1929. Until then, the majority of those guiding economic policy within the peripheral economies of the world continued to see the value of a massive system of economic interchange centered on London. As a number of scholars have earlier argued, Latin America in this period was a particularly "tenacious pole of liberal orthodoxy of the modern world economy."[20]

The middle of the nineteenth century was of key significance in the history of Brazilian modernization. Under massive pressure from Britain, Brazil finally suppressed the slave trade in 1850. In the same year, the country adopted a Commercial Code, which helped to smooth the linked paths of foreign trade and investment. The pace of growth of commercial agriculture in south-central Brazil was already accelerating, nowhere more than in the Paraíba Valley of Rio de Janeiro, already "scene of the greatest coffee production in the world."[21] It was during the years after mid-century, Richard Graham has argued, that "Brazil was decisively swept into the vortex of the international economy."[22] As a recent work has emphasized, Brazil had been "forged" as a nation by this time; the country had found a new level of political stability.[23]

All of this change, not least the greater political stability, was important for Rio Grande do Sul, as it was for the nation. The decade 1835–45 had been marked by a serious regional revolt in the south, the Farroupilha. In addition, Rio Grande do Sul had been deeply enmeshed in the political turmoil of the Río de la Plata, and there were prospects for a greater degree of harmony with that region after the removal of the great regional political leader (*caudillo*) Juan Manuel de Rosas from Buenos Aires to England in 1852.[24] General order was reestablished on the *estâncias* (ranches) of Rio Grande do Sul in that year.[25] Starting as herding in the seventeenth century in parts of what became Rio Grande do Sul, ranching had evolved into a recognizable complex of habits and institutions by 1850. Paul Vidal de la Blache's classical French human geography of the early twentieth century, which dealt so much with the cultural underpinnings of regional personality, would have found a distinctive "genre de vie" here on Brazil's southern frontier.[26] Establishing its character provides our starting point. With recovery from war, the 1850's also witnessed efforts to break away from the traditional ranching mold through diversification and intensification of production. These efforts were made with differing degrees of energy in the various sections of the southern South American ranching region. Rio Grande do Sul was not at the center of the changes, but neither was it totally immune from them.

Like "tradition," "modern" is very much a relative term. Describing the Campanha as modern by 1920 does not imply the solution of all the region's structural problems. As recently as the early 1960's lack of winter forage accounted for the death by starvation of almost 250,000 cattle, a resource then worth approximately one-ninth of the state budget.[27] But there are good reasons for choosing 1920 to close this study, beyond the important pragmatic concern that Brazil took its first agricultural census in this year.[28] By 1920, the ranchers had begun to organize and to define their sectoral interests. They were pressing for improvements on a broad front, including in such areas of weakness as the operation of the regional railway system and the port at Rio Grande, as well as the slim degree of long-term rural credit available. In addition, foreign-owned frozen-packing plants (*frigoríficos*) had been established towards the end of World War I. Though their significance in 1920 was still symbolic as much as real, it is clear that the crucial first steps in the direction of a modern form for local pastoral activities had already been taken. Contemporary observers completely understood the importance of this. Clayton Sedgwick Cooper, an American who visited the region during World War I, saw the whole of southern Brazil as standing on the threshold of rapid economic development, including in its pastoral industries:

She [Brazil] believes, and with good right to the belief, that she is to become the great cattle country of the earth; and that Rio Grande do Sul will lead and that Matto Grosso and other inland states will follow in this development. Already five big freezing plants are establishing themselves in south Brazil . . . and leading to an industry that promises to eclipse anything that Australia, Argentina or the United States have yet accomplished in this business of feeding the world.[29]

This was unabashed boosterism, to be sure, coming from a commercial source; even so, a similar optimism infuses the regional analyses of ranching by economic geographers writing at this time. For the German geographer Leo Waibel, a scholar little given to hyperbole, the rapidity of the transformation of Brazilian ranching during the First World War represented a "development without par in the world economy."[30] Most of the capital invested in Rio Grande do Sul's new freezing plants was American, and this marks another important change. In terms of the "core," there is clear evidence that the center of the world economy had shifted to the United States.

In order to begin to understand how Rio Grande do Sul fitted into the world economy, it is vital to keep the international context to the fore. The contemporary literature of the period can easily lead to a distorted perception of economic change. The problems contemporary Riograndense observers claimed were unique to their region were often in fact shared by their neighbors, so a broad canvas is needed. Enlightened ranchers from Rio Grande do Sul thinking about economic improvement tended to look beyond Brazil for an example. In the most exceptional cases, they looked far afield to parts of the English-speaking world, such as the United States or Australia for methods, or England for pedigree livestock. But this distant reach for inspiration was a rare phenomenon, and usually Riograndenses turned immediately southward. In the decades before the First World War, Argentina and Uruguay were common symbols for progress within Latin America.[31] Their roles in the transformation of the Campanha require clarification. And while they were widely used as a yardstick for measuring the progress of Riograndense development, these countries also experienced an uneven modernization.[32] There was more to Argentina than Buenos Aires Province, as scholars are increasingly rediscovering. The regional differences in modernization need uncovering, both at the level of the individual countries and at the more general scale of temperate South America.[33] It is fitting to emphasize here, as well, that southern South America formed only one of the many peripheral stages on which the lateral extension of the Industrial Revolution began to be played out in the decades preceding World War I.[34]

The scale of British commercial activity in the Río de la Plata was great enough that "honorary dominion" status has been widely conferred on Argentina and Uruguay. Just what did this label signify in terms of the geographical patterns of development that the various South American countries experienced?[35] It appears self-evident that the effects of "metropolitan influence" were not spread evenly, yet this facet of change does not always emerge very clearly in the literature. For example, Peter Winn has argued that Uruguay formed part of Britain's "informal empire" by the end of the nineteenth century, but it is clear that much of this British, as well as other foreign, interest focused almost entirely on the southern section of the country.[36] In Brazil also, many of Britain's banking or broader economic interests concentrated in the south-central parts of the nation, as Richard Graham's study of British influence showed.[37]

Taking this broad comparative approach, it quickly becomes clear that Uruguay is especially important in any attempt to establish a division between "traditional" and "modern" forms in South American ranching. Uruguayan scholars, especially the research team of José Pedro Barrán and Benjamín Nahum, have repeatedly demonstrated the weight of the Brazilian presence in the economic development of their country, a presence associated almost exclusively with traditionalism.[38]

In the "race for improvement" during the nineteenth century, the ranchers of Rio Grande do Sul, with their self-confessed apathy towards new ideas and techniques, often appeared to be the losers.[39] Yet modernization was no automatic panacea for economic weaknesses. Barbed wire radically improved the chances for stock management; it also disrupted the social structure by displacing ranch hands and many of the *agregados*, the long-established occupants of small portions of the estâncias. Pedigree animals walked impressively before judges in the expositions, such as at the famous meetings of breeders at Palermo on the outskirts of Buenos Aires, yet those same animals also had far less resistance to certain diseases and to summer droughts than the local creole breeds. In short, the thinking structures underpinning a very conservative set of rural societies were not always in agreement with those of the Europeanizing liberals. This calls for a closer examination of the nature of rural society in Rio Grande do Sul and for the identification of points of contact—and disagreement—between the outward-looking liberals and the die-hard conservatives who looked no further than the natural margins of their own properties. The ranchers of Rio Grande do Sul were by no means as irrational in their caution as some contemporary sources might lead us to consider them. In the attitudes ranchers displayed toward modernization, noneconomic factors came into play throughout the nineteenth century—and even beyond.

The development of ranching in the Campanha of Rio Grande do Sul in the period between 1850 and 1920 is also in a sense the history of a society under siege. It was under siege not so much by the more conventionally understood international political or military pressures as by the effects on prevailing ideas and value systems of the technological transformations being wrought. The modernization of ranching, epitomized by such changes as the adoption of the barbed-wire fence, marginalized yet further those gaúchos without access to their own land. What exactly did the diffusion of a modern rural apparatus mean in social terms? In Alistair Hennessy's words, the grasslands of Rio Grande do Sul "had been a disputed region of marauding cattle raiders and smugglers and tended to retain this character until the latter years of the nineteenth century."[40] This characterization of the social history of the region is partially misleading, in that it overstates the degree of social turbulence. There were armed struggles and there were marauding bands, as the reports of the province's many presidents make abundantly clear. Unemployed or underemployed labor was important in the social disorder, but it stretches the evidence to take Hennessy's "marauding cattle raiders and smugglers" as the leitmotiv for the social history of the area.[41] On the other hand, technological transformations, expressed in the landscape most dramatically through barbed-wire fencing, squeezed some of the rural labor within the pastoral societies. The "traditional" ranch relied on the available labor of occasional workers (agregados). Division of the range with wire radically changed the cycle of labor on the estâncias. It usually called for an even smaller core staff than was needed in 1850 and did away with much of the need for occasional help. At least some of the explanation for the social disruptions affecting Rio Grande do Sul and Uruguay around the end of the nineteenth century must be sought in the fact that a way of life was drawing to a close. In this context, Bradford Burns's thesis that the nineteenth century saw a growing conflict between the value systems of European-influenced "liberal modernizers" and the "folk" is significant for the interpretation of Latin American history.[42] The "traditional" gaúcho society may not have had such deep roots as is sometimes implied, but a strong argument can be advanced that northern Uruguay and Rio

Grande do Sul together constituted a functional region that was one of the last repositories within temperate South America for a certain colorful type of existence.[43]

Researchers are given to claiming the existence of neglected topics, yet the development of Rio Grande do Sul is truly one of these. By 1920, Rio Grande do Sul held third rank in Brazil in the overall value of its production, but analyses of Brazil's development largely ignore it. For example, while São Paulo, Minas Gerais, and Rio de Janeiro (the other leading states) all have multiple entries in the index to Nathaniel Leff's distinguished essays on Brazil's underdevelopment and development, Rio Grande do Sul is not even listed, despite his extended argument that the domestic economy should stand high on the research agenda.[44] Information on Rio Grande do Sul is also sketchy in Steven Topik's analysis of the political economy of the Brazilian state, another important national-scale study.[45] These omissions serve only to draw our attention to the urgent need for more research.

At the same time that existing national treatments of Brazil are recognized as deficient in their treatment of regional issues, the importance of regional economic and social issues has been increasingly emphasized.[46] There have been multiple calls for case studies of specific Brazilian systems. And within those systems, ranching is widely regarded as a particularly neglected sector.[47] Scholars argue that it has not been studied systematically or in sufficient detail, even though decades have passed since Joseph Love argued that "for the economic historian, Rio Grande do Sul is an area ripe for investigation" and further that "an economic study of the state might include comparisons with Argentine and Uruguayan patterns of development."[48]

Topics involving visible, dramatic change attract researchers in historical disciplines. In the context of southern South America, it comes as no surprise, then, that the vast agricultural transformation of the Argentinean humid pampas in the seven or eight decades after 1850 has garnered most attention. By 1920, Buenos Aires, one of the world's great cities, reached its apogee in terms of world attention largely because of these grand changes in the rural sector. Both this rural change and the growth of the city itself have been the subjects of books written with great elegance and economy of words by the late James Scobie.[49] For the core of the great pampa grassland region, there is an increasingly specialized historical research literature, but moving away from the estuary, knowledge of rural society and economy remains very general.[50] The decision to examine Rio Grande do Sul's pastoral sector reflects a conscious choice to join researchers probing the "peripheries within the periphery," which, after all, was most of Latin America in the nineteenth century.

Like all books, this study builds on earlier contributions.[51] But in the related disciplines of economic and social history, or of historical geography, there is still much to be done. The existing bibliography for the rural history and geography of Rio Grande do Sul's ranching zones resembles some of the region's breeding cattle of the nineteenth century— neither desperately thin nor especially fat. Earlier studies have tended to be either vast in temporal scope, or very specialized.[52] Efforts to build Rio Grande do Sul into an explicitly comparative framework have been few, although this is starting to change. Richard Slatta's recent bold experiment at synthesizing the various cowboy cultures of the Americas is particularly noteworthy.[53] The few existing comparative studies have invariably been made by scholars whose primary research base lies in the Plata. Not surprisingly, the value of their research conclusions about Rio Grande do Sul, a pampa periphery, sometimes seems tenuous. In particular, Slatta's assertion that Rio Grande do Sul's ranching culture represented a "static frontier" in the nineteenth century represents a conceptualization of the region that this book is bound to revise.[54]

Despite the clear importance of understanding more about the development of one of Brazil's most important regions, the present study is among the first to direct its focus to the transformation of the ranching economy in its key period. In terms of the historiography of ranching along Brazil's southern frontier, this study is distinctive for choosing to pay detailed attention to what can be considered the "neglected" nineteenth century in Rio Grande do Sul. Throughout the research for this work, a conviction has remained that the period from 1850 to 1920, the era marked by the first stirrings of a true "world economy," warrants a separate treatment.

This book explores the changes that took place in the Campanha of Rio Grande do Sul between 1850 and 1920. While change was by no means as visible here as in the neighboring countries to the south, slow development did take place. The particular character of change in this society and economy warrants its own explanation. Within the great temperate grassland region, the Campanha confronted a distinctive range of physical, social, economic, and political constraints on its development, bearing closer comparison with other parts of the South American ranching periphery, such as Mesopotamian Argentina (Frontispiece) and especially northern Uruguay.[55] The wave of "informal empire" that passed through the province of Buenos Aires was much attenuated by the time it reached Rio Grande do Sul. Among the objectives of this work are to chart that modernization ripple, to account for its pace and direction, and to identify the structural and cultural obstacles in its path.

Organized around cross-sections and following vertical themes (the changing geographies of social and economic phenomena through time), this book draws most of its obvious inspiration from what has become a classical historical geography, inspired chiefly by the work of Clifford Darby.[56] Firmly rooted in the archive, especially for the reconstruction of an earlier geography of ranching in which printed material is limited, the study joins much historical geography in having a strong empirical base. The book is organized along the following lines. Through a historical-geographical synthesis, the opening chapter provides essential regional background. Chapter 2 establishes the character of the "traditional" ranching economy around 1850; it gives an analytical portrait of the regional expression of the gaúcho system before extensive modification. With its focus on the processing and marketing of ranch production, Chapter 3 serves a similar function from the perspective of the coastal complex of factory and port. The book then fills out the theme of the changing geography of ranching over time, beginning with three chapters on the spread of innovations. Chapter 4, about the contemporary "theory" of modernization, traces the emergence of an institutional framework—the leading edge of change as expressed through agricultural societies—and surveys the tangible effects of these bodies through the emergence of publications, exhibitions, and rural legal codes. The focus is then directed more squarely toward the patterns of adoption of material phenomena. Chapter 5 examines the arrival of the leading indexes of modernization on the estâncias: improved breeds of animals, wire fencing, and planted pastures. Chapter 6 takes the same theme beyond the ranch, concentrating on the spread of innovations in the processing and marketing of meat. An explanation of the rate and pattern of modernization of the Campanha ranching economy forms the focus of Chapter 7. Like the early pair of chapters on tradition, Chapter 8 consists primarily of horizontal description. Incorporating data drawn from Brazil's first agricultural census, this brief epilogue aims to strengthen understanding of what an uneven modernization of ranching had entailed. A concluding chapter summarizes main findings and suggests themes for further research.

CHAPTER ONE

Rio Grande do Sul

BEFORE IT EMPTIES into the Río de la Plata (River Plate), the Uruguay River describes a great arc of more than 1,600 kilometers on the South American map. The land between this river and the Atlantic comprises almost all of Rio Grande do Sul and Uruguay. Since this territory also contains a major cultural boundary between Spanish and Portuguese Americas, the toponymy reflects an involved history. For example, one regional historian has investigated 41 names for the territory of Rio Grande do Sul over its four centuries of history.[1] The same topic requires some care for Uruguay. In colonial times, this area was usually known as the Banda Oriental, reflecting its situation on the east bank of the Uruguay River. When it gained its independence, it took the official title of the República Oriental del Uruguay. During the nineteenth century, it was often referred to in southern Brazil as the Estado Oriental, and its inhabitants were more likely to be described as "Orientals" than as Uruguayans.

Rio Grande do Sul (Figure 1.1) constitutes Brazil's southern frontier and has a distinctive geographical character within the nation. Extending between 27° S and 34° S, it lies below the tropics. Therein lay much of the region's appeal during the nineteenth century for groups such as European colonists and travelers. Rio Grande do Sul's climate is of the Köppen Cfa type, wet with mild winters, hot summers, and no dry season. The region receives plenty of rain, generally between 1,200 and 1,800 mm per year. Seasonality is pronounced, with temperature differences between summer and winter averaging 13°C. Frosts are not uncommon.[2] The still tropical summers permit the growth of an important series of annual crops, including maize, rice, soybeans, manioc, and tobacco. Rio Grande do Sul has had a long but problematic historical association with wheat, grown as a winter crop. Throughout southern Brazil, winter crops pose sharper agronomic challenges than summer crops, mainly on account of instability of rainfall, high humidity, and

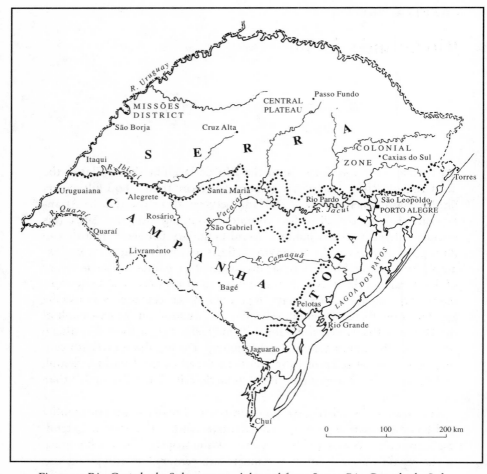

Fig. 1.1. Rio Grande do Sul, c. 1920. Adapted from Love, *Rio Grande do Sul and Brazilian Regionalism* (endpapers).

frosts.[3] During the nineteenth century, British consuls were usually quick to comment that Rio Grande do Sul was Brazil's sole major supplier of temperate products and had high hopes for it in this respect. These expectations of economic development began to be met during Brazil's Old Republic.

The economic development of the twentieth century, underpinned by commercial agriculture and the emergence of industry, has been accompanied by a significant growth in Rio Grande do Sul's population (currently over nine million). In an interesting discussion of the regionalization of what he terms the "Neo-Europes," Alfred Crosby has pointed out that Rio Grande do Sul, Uruguay, and the humid pampa of Argentina together contain the greatest concentration of population to be found in the world south of the Tropic of Capricorn.[4]

However, at the opening of this study, Rio Grande do Sul was extremely sparsely settled. An area nearly double that of England and Wales was occupied by approximately 150,000–200,000 people (the population had been placed at 134,170 in 1846).[5] By 1890, shortly after the end of the Empire, this had increased to 897,455, and it had reached 2,182,713 by the close of the study in 1920. Throughout the nineteenth century the Campanha was losing ground in terms of its share of the demographic weight, a tendency that intensified during the Old Republic. By 1920, only 20 percent of Rio Grande do Sul's population inhabited the Campanha, and even today the population density in that region remains very low, at under eleven persons per square kilometer. In contrast, the Serra, which held an insignificant proportion of the regional population at mid-century, had become home to half of Rio Grande do Sul's people by 1920.[6]

Contrary to a popular conception, Rio Grande do Sul is not a region displaying a simple geography of unbroken grasslands.[7] There are marked regional contrasts in the natural vegetation (Figure 1.2), closely related to the relief. Based on the relief, it has been customary to divide the state into six or more physical regions.[8] Three of the major physiographic divisions are particularly important for interpreting the regional human geography: the Litoral, the Serra, and the Campanha.

The Litoral, or coastal district, is composed of poor, sandy soils in the main. Despite its environmental limitations, it has played an important part in the economic development of Rio Grande do Sul. It was settled early within local colonial history (the garrison at Rio Grande was first established in 1737, for example), and the great lagoons of Mirim and Patos have played vital roles in fostering fluvial communications between the interior districts and the Atlantic.

In much of the northern part of the state, the Serra, the landforms are dominated by an extension of the Paraná Plateau, which provides a hilly relief similar to that found in other sections of southern Brazil. Much of the Serra is composed of the Planalto, a tableland inclined toward the west. In the east the Planalto averages 800–1,000 meters in elevation, forming a very steep break with the Atlantic; in the west, however, it

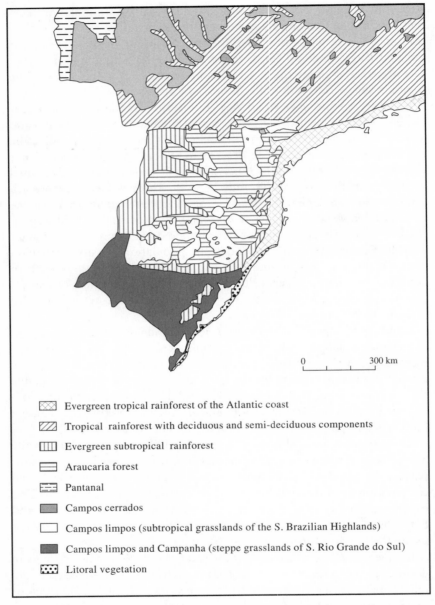

0   300 km

⊠ Evergreen tropical rainforest of the Atlantic coast

▨ Tropical rainforest with deciduous and semi-deciduous components

▥ Evergreen subtropical rainforest

▤ Araucaria forest

▤ Pantanal

▨ Campos cerrados

☐ Campos limpos (subtropical grasslands of the S. Brazilian Highlands)

▨ Campos limpos and Campanha (steppe grasslands of S. Rio Grande do Sul)

▨ Litoral vegetation

Fig. 1.2. The natural vegetation of southern Brazil. After map in Pfeifer, "Kontraste in Rio Grande do Sul," p. 279.

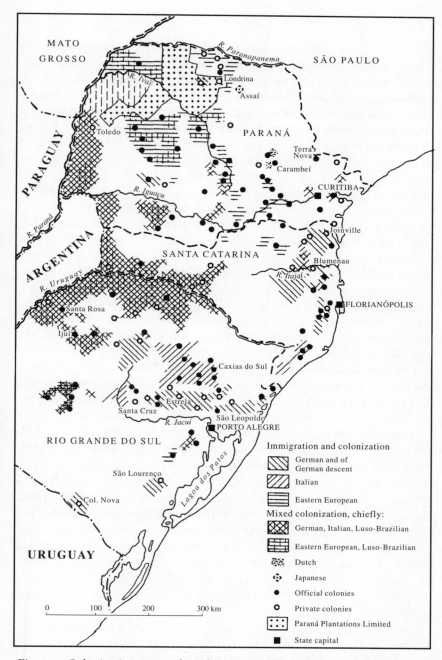

Fig. 1.3. Colonization zones of southern Brazil. After map in Kohlhepp, "Donauschwaben in Brasilien," p. 363.

falls gently from 300 to 100 meters. The natural vegetation of the highest portions of the Planalto is *campo limpo*, subtropical grassland generally clear of trees. At the opening of this study, forest still covered much of the Serra. However, the work of a large-scale European colonization with fire and ax was beginning to make its mark, transforming an extensive region into one of family-based agriculture on small farms.[9] German settlement began in 1824 at São Leopoldo, between the Jacuí River and the Serra escarpment. It continued in a zone about 200 kilometers to the west, later filling the whole subtropical forest fringe (Figure 1.3). In Rio Grande do Sul, German settlement generally stayed below 500–600 meters, the lower limit of the Paraná pine (*Araucaria angustifolia*), a pattern governed by the colonists' perception of the fertility of the soils. The Italians, who began to arrive after 1870, settled on higher ground. Beginning in the 1890's, the extension of railways across the Planalto took colonists, by now of much broader origins, into further untapped zones of virgin forest. Figure 1.3 shows the major colonial zones of southern Brazil. A comparison of Figures 1.2 and 1.3 reveals a cardinal point in the geography of southern Brazil: almost all of the European colonization occurred on forested lands, reflecting the generally higher initial fertility of their soils.

The final major subregion is the Campanha. This is the zone of greatest direct significance for the study of ranching in Rio Grande do Sul, and there is no good analog region for it elsewhere in Brazil. The inhabitants of Rio Grande do Sul understand different things by the word "Campanha." In a physiographic sense, it comprises the region of rolling hills (*coxilhas*) and uninterrupted steppe-like grasslands that extends from Bagé to Uruguaiana in the southwest.[10] A broader regionalization is based on the areas that have displayed marked similarities in terms of historical processes. In this case, the label "Campanha" can be fitted comfortably to the whole of the southern half of Rio Grande do Sul, which lies inland from the western fringes of the Patos and Mirim lagoons. This broader regionalization is the one usually adopted in this work. However, the matter could be loosened even further. For example, for the compilers of regional vocabulary, "Campanha" stands as a generic term for all of the rural areas, the zones beyond the major towns of the Litoral.

The southern portion of Rio Grande do Sul is not uniform with respect to relief, and the "flat as a board" image drawn from much of the pampa of Buenos Aires does not serve here. In the southwest, the Campanha proper, the coxilhas average some 200 meters in height. Between

Fig. 1.4. Grasslands of the Campanha near Bagé, c. 1920. From Brazil, *Recenseamento de 1920*, 3(1), following p. 385.

this region and the sandy Litoral there is a distinctive subregion known as the Serra do Sudeste. Here the relief is much more irregular and reaches 500 meters at times.[11] In addition, the soil geography of the region is complex (and still incompletely understood in its details), reflecting in part the differential erosion of a wide variety of rocks, many of them metamorphic and sedimentary in character. Brunizem and reddish prairie soils predominate.[12] The Campanha west of a line drawn roughly from Santana do Livramento (Livramento) to Alegrete has a distinctive physical base; its geomorphology is dominated by an extensively eroded tableland (*planalto*) of igneous rock, gently inclined towards the Uruguay River. Basalt, the same material that is responsible for much of the highlands of the Serra, underlies the soils here. The influence of the bedrock is much more visible in the Campanha than on the coastal plains of Argentina, where it lies obscured by vast deposits of sand and clay.[13] While the Campanha's soils fall into the naturally fertile category, complaints about their weak mineral content, especially of phosphates, have been common.[14] This is undoubtedly part of the reason why the Campanha was slow to attract large-scale cultivation.[15]

Although it comprises only a small portion of Brazil's grassland area, the Campanha contains by far the best natural pastures in this vast country. Grass distributions are always in flux, not least as ranchers search for more successful forage; like the pampas, the Campanha has seen great change in its flora as a result of exotic invasions.[16] Based on extensive direct observation of the land in 1820–21, the French traveler Auguste de Saint-Hilaire, a distinguished botanist, came to appreciate that Rio Grande do Sul was a transition zone. The quality of the pastures was at its highest for ranching in the Campanha, especially in the Rincão da Cruz, an area east of Itaqui that contained around thirty estâncias. Even here, the pastures were not the delicate grasses of the Montevideo region, which Saint-Hilaire described as of "such a quality that they should never serve as a term of comparison," an observation often forgotten in subsequent times. On the other hand, while the vegetation of the Campanha lacked the "happy verdure" of southern Uruguay, it was "infinitely more varied," and the grasslands resembled "a vast garden."[17] By the early decades of the twentieth century, short, perennial grasses formed most of the plant cover of the Campanha. Roseveare's study of Latin American grasslands (based mainly on a review of interwar research) mentions specifically the genera *Paspalum* and *Axonopus*, as well as legumes of the genera *Trifolium* and *Adesmia*.[18] The leading distinction the local inhabitants make in the quality of the natural pastures is between *campo fino* and *campo grosso*, not far from what Darwin saw in 1833 as "a carpet of fine green verdure" and "coarse herbage."[19] The core area of the campo fino is the Campanha along the Uruguayan frontier from Bagé to Uruguaiana (Figure 1.1). From field observations made in 1965, the cultural geographer Gottfried Pfeifer noted that the traveler encountered the true campo grosso, subtropical pasture, north of a line drawn roughly from Santa Maria to São Borja. Between these two extremes, there is an extensive transition zone, where weeds and trailing shrubs (generally described as *chirca*) have invaded the pastures. Such invaded pastures are commonly known as *macegais*.[20] Toward the end of the period under study, the authorities were unable to be categorical about the regional delimitations of grasses; even the best Campanha grasslands close to the Uruguayan frontier still contained stretches of low nutritive value.[21] Tall-grass formations, such as *Aristida pallens* (approximately 50 cm in height and of impoverished aspect), offered no forage in autumn.[22] But ranchers have tolerated this and similar taller species on account of the protection they have offered their livestock from the winter winds.

While temperate, the aspect of the land shows some remarkable changes through the seasons. It is the damp winters which usually occasion some surprise in visitors. The Conde d'Eu, son-in-law of the Emperor, summarized this well in the 1860's, pointing out that the soils and vegetation of the Campanha wore a boreal aspect in the full of winter, similar in his view to the moors of Britain.[23] Rio Grande do Sul has very much a character of transition between the major South American physical regions. This has led to difficulties of classification at times.[24] However, most of the grassland portions of the region lean physiographically towards the Plata and are seen as having more in common with parts of that region than with the rest of Brazil.

Geopolitics and the Grasslands

Rio Grande do Sul became a part of Brazil as a result of human initiatives.[25] Its territory fell within the Spanish sphere theoretically, but the two powers were unable to decide where exactly the diplomatic boundary fostered by the Treaty of Tordesillas (1494) was located. Thus the history of its incorporation into Brazilian territory is basically the expression of Portuguese expansionism, a "Drang nach Süden" in Dauril Alden's phrasing, which unfolded in uneven stages.[26] For centuries the two Iberian metropolitan powers were engaged in an extended struggle for the control of space, and southern Brazil stood at the center of any potential conflict.[27] One important guide to Portuguese ambition was the conviction that the Plata formed a natural boundary to their dominion. At the same time, the struggle also developed into conflict over the exploitation of a mobile economic resource, namely cattle.

Within the modern-day territory of Rio Grande do Sul, the settlement chronology of European expansion begins with the Spanish Jesuits of the famous Paraguayan missions, who were to have a long yet checkered history of activity in the region. The Jesuits came east of the Uruguay River into lands occupied by Tape Indians in 1619. During the following decades they established an impressive complex of settlements (reductions) in what is now Rio Grande do Sul. By the early 1630's, they had pushed eastward almost as far as the Atlantic coast. This expansion did not last, and by the end of the 1630's some of these missions already lay in ruins.[28] This was a result of attacks mounted by raiders from the São Paulo area, the well-known *bandeirantes* ("pathfinders"), seeking to

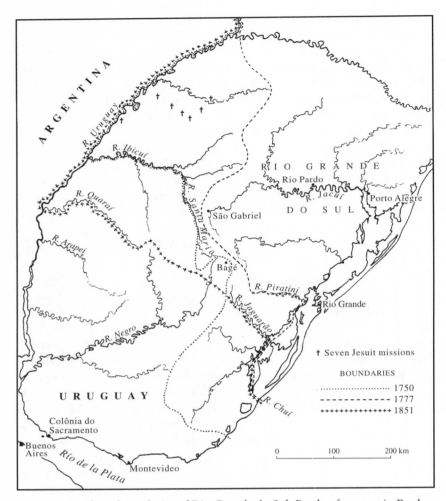

Fig. 1.5. Southern boundaries of Rio Grande do Sul. Partly after map in Roche, *Colonização alemã e o Rio Grande do Sul*, 1: 12.

enslave the docile Indian labor under the tutelage of the Jesuits. Facing mounting attacks on their reductions, the Jesuits withdrew from the territory of Rio Grande do Sul around mid-century. A persistent group, they came back across the Uruguay River in the 1680's and founded seven important missions, the so-called Sete Povos. With full consideration given to defense, these were closely grouped and deliberately placed not far from the main river.[29] The whole territory around the seven missions became known as Missões, following the character of the settlements (Figure 1.5).

Early in the organization of their reductions, the Jesuits had introduced livestock into the pastures. They were to wield a profound influence on the regional geography. The subsequent dispersion of cattle throughout the southern region was largely a result of the disruption of the Jesuit mission economies by the bandeirantes. The Jesuits themselves learned from experience the value of allowing animals to propagate over wide expanses of grassland (*vacarias*).[30] Frequent disruptions to the working of the Jesuit rural economy during the seventeenth century only served to allow livestock to disperse over ever broader areas. While the economic rationale for mission raids had at first been manhunting, the raiders who descended from the plateau quickly grew aware of what a considerable potential resource the feral and semi-feral cattle represented (initially for their hides and tallow much more than for any meat).

Even though the sphere of Jesuit activities came to extend well beyond Missões, the lands of what would become Rio Grande do Sul were largely empty of people towards the end of the seventeenth century. As the Portuguese prepared to take their bold step of founding the colony of Sacramento (1680) at the mouth of the Plata estuary, south of Laguna in Santa Catarina a true no-man's-land (*terra de ninguém*) extended to the Plata, open to the territorial ambitions of both Spain and Portugal. The reasons for the foundation of Sacramento were various, but among them was Portugal's will to reinforce its claims to the vast grazing lands known as the *vacarias do mar*.

While the temperate grasslands of southern Brazil offered the most favorable natural environment for the settlement of Europeans in the whole of the colony, Portuguese interest in them did not always show an even quality. Hides, the major product taken from the wild cattle ranging on these grasslands, formed one of Brazil's staple exports throughout the colonial period, but they occasioned far less excitement for those who guided the fortunes of the Portuguese royal coffers than did gold.[31] However, interest in the south grew during the eighteenth century. An important trade in mules developed to serve the mining economy of south-central Brazil. Clearly the militarized settlement at Sacramento also had an economic basis. For example, during 1726–34, a phase when the economy of the settlement was particularly sound, that colony was exporting between 400,000 and 500,000 hides per year. As Guilhermino Cesar has wisely pointed out, such elevated figures could only have been achieved through an extremely assiduous Portuguese collaboration with the Spaniards.[32]

The next decisive step in the Portuguese expansion was the foundation of a garrison-settlement at Rio Grande in 1737. By this time, informed opinion was swinging very much in favor of making serious efforts to retain as much as possible in the south. André Ribeiro Coutinho, a Portuguese soldier with extensive experience in foreign service behind him, praised the lands of Rio Grande in the same year as a region of abundance. In the process, he provided a fine lyrical description of the characteristics of a zone of transition between subtropical and temperate environments:

> For here is plenty of meat, plenty of fish, plenty of geese, plenty of wild duck, plenty of kingfishers, plenty of partridges, plenty of *jacum*, plenty of milk and cheese, plenty of pineapples, plenty of hides, plenty of timber, plenty of clay, plenty of balsam, plenty of hills, plenty of lakes and plenty of marshes. In the summer, plenty of heat, plenty of flies, plenty of *motuca*, plenty of mosquitoes, plenty of moths, plenty of fleas. In the winter, plenty of rain, plenty of wind, plenty of cold, plenty of thunder; and all the year round, plenty of work, plenty of making fascines, plenty of excellent air, plenty of good water, plenty of hope, and plenty of health.[33]

As the century drew on, it was evident that Portuguese royal interest was stronger in the development of the southern borderlands than it was in the economic integration of the interior of Brazil's vast territory.[34]

Efforts to delineate Spanish and Portuguese territories in the region punctuated the eighteenth century. Attempts to trace boundaries were largely reflections of the metropolitan rivalries in Europe. The most important forays into definition came with the Treaties of Madrid (1750) and San Ildefonso (1777). Under the provisions of the former, the middle Uruguay River formed the boundary between the Spanish and Portuguese possessions. Rather than see the Spanish Jesuits of the Sete Povos isolated in the new Brazil, the Treaty of Madrid called for the removal of their missions into Spanish territory west of the Uruguay River. Neat in paper convention, the planned removal of the Sete Povos became a brutal affair. The Indians put up a valiant resistance to the Portuguese expansion. Rightly or wrongly, the Jesuits were seen to be fomenting this resistance and ran afoul of both the Portuguese and the Spanish as a result. This was a leading excuse behind their expulsion from Brazil in 1760 and from Spanish America in 1767.[35]

The Jesuits had been a complicating element in South American geo-politics, but even with their removal Spain and Portugal were still fighting over the grasslands. Under the Treaty of San Ildefonso, the Portuguese waived their right to hold the colony of Sacramento and in return regained their settlement at Rio Grande, captured by the Spanish some years earlier. On balance, this treaty was territorially unfavorable to the Portuguese, not least in that the Spanish regained a presence in Missões. On the other hand, as the Spanish polymath Félix de Azara made clear in his famous manuscript "Memoria rural del Río de la Plata," Portuguese-speaking ranchers took little notice of paper treaties and continued to press forward "day and night."[36] The 1801 war in Europe between Carlos IV and Portugal, while short, gave the local military command in Rio Grande do Sul the chance for a conquest of the former Jesuit missions along the eastern bank of the Uruguay River.[37] With the incorporation of Missões into its territory, Rio Grande do Sul took on roughly its present form.

By the end of the colonial period, a ranching pole had developed that included more than the lands of Rio Grande do Sul itself. On a frontier that had witnessed numerous conflicts, of various durations, a cultural region had evolved where certain groups had become accustomed to profiting from the spoils of war. Even after the incorporation of Missões into Portuguese territory, some Brazilian military officials continued to head down into the Banda Oriental in search of livestock. Most of these armed incursions came from the area between the Ibicuí and Quaraí Rivers, lands that were not yet fully integrated into Brazilian territory and that were populated mainly by transient elements of both nationalities.[38] Since many of the soldiers in the frontier campaigns were irregulars, they found their payment in the form of livestock sacked from the enemy. By the nineteenth century, this matter of finding personal and local solutions to liquidity problems had evolved into a habit of mind. It was to complicate greatly Brazil's relations with its neighbors.

While the territory controlled by the Portuguese on the southern frontier after 1801 already resembled very closely the shape of modern Rio Grande do Sul, this did not denote an end to military disturbances. The Viceroyalty of the Río de la Plata, created in 1776 with its center at Buenos Aires, began to fall apart after 1810. Revolution did not lead to the smooth formation of nation-states around the Plata, and near-endless warring between rival groups for the exercise of political power marked the first half of the century. The struggle for Uruguay has been usefully oversimplified as one "between an urban literary elite, attempting to impose something like European political norms, and caudillos with rural ties and little sense of constitutionalism."[39] While the transfer from colony to nation was vastly less complicated for Brazil, the insecurities occasioned by *caudillismo* to the south affected Rio Grande do Sul. For example, frontiers remained undefined. The important boundary with Uruguay was not traced until 1851. Nor did Brazil's territorial ambitions for more temperate grasslands cease as the colonial period drew to its close.

For example, the flight of the Portuguese royal family and their court to Rio de Janeiro in 1808 was accompanied by territorial ambitions in the south. In the Banda Oriental, there was a rural insurrection under the leadership of José Artigas after 1811. This drew the Portuguese, under the pretext of wishing to help the Spanish royalist cause, into the area. They invaded the Banda Oriental in 1816 and annexed it to Brazil as the Cisplatine Province in 1820–21.[40] Some looked favorably on their occupation. Thomas Hood, the British consul in Montevideo, thought that Brazilian settlement of the interior districts would lead to political stability.[41] The vision that Hood forwarded to London was illusory. Juan Antonio Lavalleja invaded the Banda Oriental from Argentina in 1825 and fought a war with Brazil that ended in 1828 with the establishment of the independent state of Uruguay, brought into being with the aid of considerable British pressure.[42] Until the technical means became available for Montevideo to extend its centralizing influence toward the Brazilian frontier, the political character of this new state was for decades to be one of chronic instability. Conflict in Argentina about what form the state should adopt spilled over the Plata estuary into Uruguay. Uruguay was the scene of civil war through the entire decade of the 1840's (the Guerra Grande of 1839–51). Ranchers from Rio Grande do Sul were easily drawn into the conflict, not least to protect their ranching interests south of the frontier.[43]

In addition, the southern flank of Brazil made its own protest against the central authority of Rio de Janeiro in this period. The conduct of military affairs was one element that built a strong sense of regional grievance. Between 1825 and 1828, when Brazil was waging war with the United Provinces (Argentina) over the territory of the Banda Oriental, soldiers from Rio Grande do Sul made up the bulk of the army. Pay in arrears, inadequate supplies, and allegedly ineffective command from outside combined to seed a growing disdain for central authority. In addition, the loss of the Banda Oriental in 1828 did not sit well within Rio Grande do Sul. Thus, politico-economic and military complaints were already carefully nurtured by some in Rio Grande do Sul even before the Regency government of 1831 exacerbated the situation.

After 1835, parts of the Campanha formed a theater of war for almost a decade as the base of operation for those involved in the Farroupilha Revolt (1835–45). The root cause of the war was intolerance for inept central authority at a time of slower regional economic growth following a phase of several decades of impressive organization in the pastoral economy. In this vein, Spencer Leitman, the leading modern student of the event, has argued that it "can be considered not merely as the arbitrary response of one caudillo [Bento Gonçalves da Silva], but as a reaction of a socioeconomic group [the ranchers of the Campanha] to the obstacles of economic growth."[44] Historians have argued at length about the rising. It was certainly the most serious threat to its territorial integrity that the Brazilian Empire faced.[45] It was also of vast importance in the creation of Rio Grande do Sul's regional identity. The Farroupilha left a legacy of incomplete understanding between Rio Grande do Sul and the remainder of Brazil. The instabilities that it occasioned, along with the broader regional disputes of the first half of the century, were clearly no basis for rural modernization.

Land and Settlement

During these centuries of military disturbance, the land of the region was taken up. In order to hold territory, it was obviously in the Portuguese strategic interest to occupy it. Land grants conferred by royal authority formed the basic official means used to organize rural space and to foster settlement. The types employed in Rio Grande do Sul, the *sesmaria* and the *data*, were distinguished from each other by a vast difference in size. Sesmaria concessions comprised large tracts of land, usually of up to three square leagues (13,068 hectares) in area. Such areas of land lent themselves to extensive ranching, and their concessionaires soon established estâncias within them.[46]

There were efforts in the colonial period to intensify the occupation of space in various ways. For example, occasional initiatives were taken to check legal abuses in the awarding of land, and officially sponsored efforts were made to foster the cultivation of specialized agricultural products. Little came of these. Around the middle of the eighteenth century, there was a significant effort made to settle poor Azorian families as farmers by giving them datas (of approximately 272 hectares), mainly in the Litoral and parts of the Central Depression (the axis of the

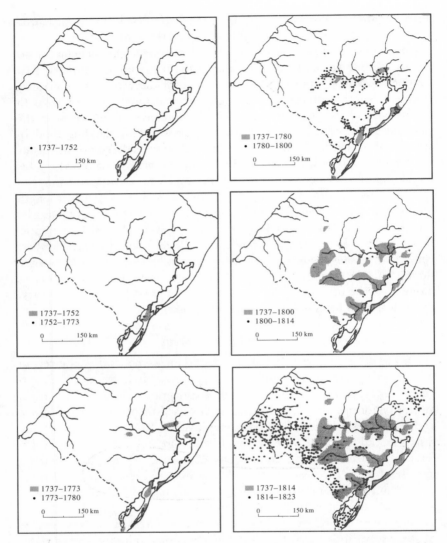

Fig. 1.6. The distribution of sesmaria grants in Rio Grande do Sul. From Lobb, "*Sesmaria* in Rio Grande do Sul," p. 57.

Jacuí Valley). While wheat cultivation flourished at the end of the century, the farming experiment did not lead far. Various reasons have been advanced for this, but the character of the cultural milieu is surely a leading factor. Extensive ranching made for a more exciting life than the backbreaking labor of growing annual crops for uncertain markets. The qualitative importance of the Azorians, in terms of their influence on the regional character, is more difficult to assess.[47] The fact remains that the vast majority of the grassland zones were conceded through the medium of the sesmarias. Before the system terminated in 1822, almost a thousand ranching sesmarias were granted and officially registered in Rio Grande do Sul.[48]

The pattern of the sesmaria grants reflects to a large extent the complicated spatial seesaw between the contending Iberian powers in the region. Thus the earliest concessions came around the garrisons in the Litoral, but from the 1750's references begin to appear in documents to the "Rio Pardo frontier."[49] The Jacuí Valley formed the major axis of entry for many of those who took up land in the Campanha proper. Gary Lobb's compilation of the sesmaria petitions has resulted in an important map-series. This has illustrated two points of especial importance (Figure 1.6). It shows that the official occupation of the land only gathered real momentum after 1780. According to the declarations of the occupants, under a quarter of the area of modern-day Rio Grande do Sul was occupied in 1785.[50] The series also demonstrates that the process of conferring sesmarias was extremely rapid in the early nineteenth century, the period during which most of the land in the Campanha was granted. The Portuguese official settlement of the Campanha, the ranching zone par excellence in Rio Grande do Sul, was an early-nineteenth-century phenomenon; in the short period from 1814 to 1817 alone, over four hundred sesmarias were granted in this region.

Lobb has characterized the sesmarias as a "successful frontier institution."[51] No doubt the prospect of gaining title to large pieces of grassland, serving as home to large numbers of feral animals, acted as a powerful stimulus for the attraction of settlers to a highly turbulent region. However, the extensive ranching system that developed was not solely the outgrowth of an institution, the sesmaria. Given the presence of feral cattle, as well as undefined (and generally undefinable) boundaries between what was Portuguese and what was Spanish, ranching was usually taking place, albeit in a most rudimentary form, before the sesmarias in a given area were actually granted. Designs of government and actual practice frequently did not mesh easily in this frontier region.

Often, the act of conferring a sesmaria represented nothing more than royal authority trying to keep pace (or even to catch up) with the status quo in a given zone.[52]

Behind the military campaigns of Portuguese expansionism lay land hunger. As early as the 1770's, before the occupation of most of modern Rio Grande do Sul, and despite the fact that the population density in the vast territory of the interior was very low indeed, within the zones understood to be Portuguese, there was the paradox that no land was available to which somebody did not already claim ownership. The military engineer Francisco João Roscio explained the basis of the land hunger that existed in the following terms: "Each inhabitant does not content himself with a few leagues of land, understanding that they are all needed by him, even though he uses only an insignificant part close to his cabin. Thus it is that the campanha is totally deserted while the lands are all given and have ownership."[53] Social equity criteria counted for little in the distribution of lands. Manoel Antônio de Magalhães perceived this, claiming in 1808 that some families were taking out three or four sesmarias in the names of their relatives while others did not hold a single palm of land. He saw this as most damaging, not only to the royal authority "but to the people in general."[54] For Samuel Gottfried Kerst, a German soldier-geographer who witnessed rural problems at close quarters during the later 1820's, land in Rio Grande do Sul had much more to do with "money and official favoritism" (*Geld und Gunst*) than any regulated processes.[55]

Recent research has confirmed the impressions conveyed by these contemporary authors. In his detailed study of the Jaguarão frontier, Sérgio da Costa Franco has demonstrated that the Portuguese Crown was already granting sesmaria land grants in the "neutral" territory between the Piratini and Jaguarão Rivers after 1789, thereby lending de jure recognition to something that was already taking place, the Luso-Brazilian occupation of the land.[56]

By making his focus the institution of the sesmaria, it appears that Lobb has exaggerated the degree of royal control over how the regional economy might have unfolded in the south and has lost sight of some of the peculiarities of the cultural geography of the region. It can be argued with some force that the cultural milieu, which Lobb traces to such a great extent from the grants of sesmarias, owed much more until the beginning of the nineteenth century to the presence, and Portuguese support, of contraband.[57]

By the eve of Brazilian Independence, the grasslands had all found owners (and sometimes several, in dispute). The case differed from that of Buenos Aires in that there were no extensive frontiers to roll back or margins to occupy. Ranching was already taking place within a relatively closed system in the Campanha. The only obvious escape from this lay to press forward into the politically unstable but emptier lands of Uruguay to the south.

The Formation of an Organized Rural Economy

Relative to other parts of the colony, Rio Grande do Sul had a very young organized economy as Brazil gained its independence, a phenomenon perhaps obscured by a literature devoted to the "heroic phases" of Riograndense development. At the same time, the integration of the Campanha into the Atlantic economy had begun very early under the military-pastoral cycle of the colonial period with the indiscriminate hunting down of animals for their hides. Even with the utterly unsophisticated means of transport available, these appear to have found their way to the coast with regularity.

By the start of the last third of the eighteenth century, another item, salt-beef (*charque*), was added to the list of pastoral exports. It was to dominate the industrial geography of the region for the next century and beyond. The origins of salt-beef processing on an industrial scale are usually ascribed to José Pinto Martins, formerly of Ceará in northeast Brazil. He had been making salt-beef (*carne de sertão*) within the northeast itself but came south in response to a catastrophic drought of 1774–77. Martins set himself up around what later became the town of Pelotas in order to capitalize on the temporary difficulties of food supply in northeast Brazil.[58] Martins was probably the first to export charque to the plantations of the northeast, but the origins of dried or salted beef in southern Brazil are rather more remote.[59] Antônio Carlos Machado credited the Jesuits with the region's first efforts at making preserved meat, as with much else on the southern frontier of Brazil. Knowledge of the basic processes then spread to the muleteers who during the early eighteenth century descended from São Paulo and Laguna on Rio Grande do Sul, in search of beasts to sell in the mining zones of south-central Brazil.

This broadening of the sources of rural wealth carried important consequences for the region. Even if the conditions of the Treaty of San Ildefonso checked the Portuguese territorial presence in the south, there is clear evidence that the economy of the region prospered after 1780, a temporary peace dividend for Rio Grande do Sul. Also, the establishment of neutral grounds between the Spanish and the Portuguese served mainly to foment contraband. As Cesar has expressed this, "the *campanha* rio-grandense . . . did not passively accept conventions dictated from Europe in its economic life."[60] In fact, there was an extensive zone of friction between the two peoples, and this did not have any clear boundary, so that it was difficult to establish at the local scale just what was Spanish and what was Portuguese. In Cesar's words, the treaty "activated the contraband of cattle as never before and in alarming proportions."[61]

By the early nineteenth century, apart from finding its obvious market among the slaves and poorer classes of Brazil, charque had gained other customers, as John Mawe revealed in an interesting discussion of what he termed "the staple productions of Rio Grande." Charque was in use for victualing ships. It had penetrated the West Indian market and had been in considerable demand there during the Napoleonic Wars. The British, during their occupation of Montevideo in 1807, had contracted for "large quantities" of salt-beef. In addition, Mawe described the quantities of hides shipped from Rio Grande (300,000 per year on average) as "almost incredible."[62]

Based on the measure of per capita exports and imports, Bauss has argued that the extreme south was the fastest-growing section of Brazil between 1780 and 1820; by the end of the colonial period, Rio Grande do Sul was the "key internal supplier of foodstuffs" to the Brazilian economy.[63] When making any assessment of very rapid economic growth, it is important to consider the type of resource base the region exhibited. Despite the short-lived florescence of a wheat trade from Rio Grande do Sul in the late colonial period, the region depended in essence on the exploitation of a staple product, the feral cattle, that existed regardless of whoever controlled it. Once political incorporation of the territory became a fact, it was not surprising that subsequent economic development was very rapid. In addition, the years of occupation and annexation of the Banda Oriental were largely beneficial for the ranching industry of Rio Grande do Sul, especially for those in the Litoral who processed the animals coming across the frontier.[64] Brazil's loss of the Cisplatine checked a long expansionary phase. The struggle for the independence of Uruguay has been linked with economic recession in Rio Grande do Sul, as well as to "the atmosphere of ill-being and of discontent that [greatly contributed] to the explosion of the Farroupilha movement in 1835."[65]

Within the interior of Rio Grande do Sul it was the Campanha that bore the strongest human imprint on the landscape for much of the nineteenth century. In their push against the Spanish, pioneers had followed the spines of the coxilhas into the interior. Almost all the towns of any significance in the Campanha had origins linked to military camps. Behind the dragoons followed merchants, contrabandists, and, shortly after, petitioners for sesmarias. By 1820, the socioeconomic matrix was virtually in place in the south. Despite the political turmoil in the region that existed for lengthy portions of the first half of the nineteenth century, the ranchers of the Campanha had devoted their efforts to creating wealth. From their work, a distinctive rural culture emerged, one which left an enduring legacy, widely known today as "tradition."

CHAPTER TWO

Gaúchos and Their Grasslands

BY THE 1850'S, the indiscriminate hunting down of cattle had
given way to organized estâncias. Within these, the gaúchos displayed an extremely strong attachment to their land and to livestock
breeding, especially of cattle, horses, and mules. The famous soldier
General Osório, a rancher himself, provided a good summary rationale
for their estância life in a reported conversation about the province with
Dom Pedro II (probably held during the Emperor's visit to Rio Grande
do Sul of 1865): "What is the best business here?" asked the Emperor.

"Raising cattle on good land, your Majesty," Osório responded.

"And the second best?"

"Raising cattle on reasonably good land."

"And the third?"

"Raising cattle on bad land," concluded the general.[1]

Osório was correct in drawing attention to an affinity for ranching
that was cultural as well as practical. But his description reduced the
workings of the regional economy to a false simplicity and provided little sense of how a narrow economy based so greatly on a single activity
actually worked in its "traditional" condition. Using sources mainly
drawn from the decade, this chapter establishes the character of the region during the 1850's. While its purpose is to provide a cross-sectional
description, the analysis of sources is not rigidly static and moves where
appropriate from the opening of the nineteenth century to around 1860.
These later years saw a crisis of success (overproduction), given the
technical means employed at the time. During that crisis, some of the
limitations of "tradition" emerged more clearly.

Estância Lands

The single aspect of the ranches that made the most powerful impression on visitors was their size. John Luccock, an English traveler with
almost exclusively mercantile preoccupations, expressed his astonish-

ment at the extent of many of the estates. After describing a property of some ten square leagues in the Pelotas region, Luccock seems to have doubted that his readers would believe him. Thus he noted that "the reported extent of farms in this part of the American Continent can scarcely be mentioned with boldness, by one who has himself little doubt of the truth of the accounts. The smallest are stated at four square leagues, or more than twenty thousand acres; the largest are said to reach a hundred square leagues, or near six hundred thousand acres."[2] Other observers bore out this judgment. In his account of the condition of ranching in Rio Grande do Sul during the early 1840's, the Visconde de São Leopoldo warned his reader not to suppose that an estância would have an area of less than two sesmarias, or six square leagues, even though the sesmaria law had limited such grants to three square leagues. The ranchers clearly sought to control the greatest area of land possible in order to increase profits and to free themselves from what a French observer called "all inopportune competition."[3] *Estancieiros* had extended their holdings either through purchase or by methods vaguely encompassed by the Visconde de São Leopoldo in the phrase "some other means."[4] It was not extraordinary, that author noted, for the ranchers to have ten or twenty square leagues of land, held conjunctly or separately.

The largest estate during the imperial period belonged to the Portuguese-born Visconde de Serro Azul, whose properties in Rio Grande do Sul and Uruguay comprised an area of some 110 square leagues at his death in 1847.[5] Expressed in the areal unit of the English landed aristocracy, this was over a million acres. Property owned on this scale placed the Visconde de Serro Azul in the same category of rural wealth as the leading landowners of Buenos Aires province. Around 1830, only the Anchorena family had more land in Argentina than the Visconde de Serro Azul held in Rio Grande do Sul and Uruguay. By the 1840's, however, the favoritism of Rosas, Argentina's dictator, and expanding frontiers of settlement in the province of Buenos Aires, had allowed for the massive expansion of the holdings of the rural elite there. The Anchorenas, for example, had increased the area of the land they controlled to 306 square leagues.[6]

At mid-century, wars against Indians were still rolling back the frontiers in Buenos Aires, adding thousands of square leagues of new grasslands that established ranching interests could incorporate into their operations. By contrast, Rio Grande do Sul in itself was a much more closed system (all of the land in the Campanha had already been taken up for decades). Given their need for more land, the estancieiros of the Campanha extended their interests into neighboring regions, such as Mesopotamian Argentina (Corrientes and Entre Ríos) but especially into Uruguay.[7] Despite the existence of some massive properties, the relative levels of landed wealth within Rio Grande do Sul appear to have been rather more homogeneous than in Argentina during the nineteenth century.

The nature of landholding in nineteenth-century Rio Grande do Sul should be a focus for further detailed research. While, in common with the Río de la Plata, some of the wealthiest families managed to add to their properties from generation to generation, the partible inheritance system was already leading to a diminution in the scale of most of the largest ranches by mid-century.[8] Drawing on information on land areas contained within an ecclesiastical survey conducted during the 1850's, John Chasteen has pointed up important regional differences in the scale of rural properties, with a clear east-west gradient in that scale (Table 2.1). Dividing the properties into four size categories, the survey data reveal the presence of small and middle-sized landholdings across the province. It is with his "large properties" of more than seven thousand hectares that the trend towards larger holdings in the Campanha proper is most clearly revealed. Based on the data presented in Table 2.1, only 5 or 6 percent of the properties fell into the "large" category in the Arroio Grande and Herval districts; moving westward, the proportion rose to a peak of 33 percent at Uruguaiana. Chasteen argues that the smaller sizes of the ranches closer to the coast during the 1850's are a reflection of longer-established settlement in those regions. A longer history of inheritance had led to a greater degree of subdivision of properties. He has a strong case. The fact that the estâncias in the southwestern areas of the province were larger may also reflect other factors. The lands of the Campanha proper form the most favorable natural environments for ranching within Rio Grande do Sul.[9] Families with commercial capital to invest tended to seek out the large properties of the interior.[10]

Except for "islands" of crop cultivation, most of the land in the Campanha and the Litoral, as well as large sections of the Serra, was devoted to the grazing of animals on open pasture. During the late colonial period, Rio Grande do Sul had developed a reputation as "the granary of Brazil" on account of its wheat cultivation but by 1850 cultivation was in retreat in the Campanha.[11]

In the 1770's, Roscio, a military engineer with a keen eye for geographical description, attributed the small scale of wheat cultivation in

TABLE 2.1

Division of Land in Selected Districts of
Rio Grande do Sul, 1854–59

District	Small properties 100–1,500 (ha)		Middle-sized properties 1,501–7,000 (ha)	
	Number	Average size	Number	Average size
Arroio Grande	53	532	38	3,296
Herval	61	696	46	3,175
São Gabriel	100	755	124	3,509
Alegrete	104	612	128	3,416
Uruguaiana	15	953	50	3,315

District	Large properties 7,001–26,500 (ha)		Giant properties 26,500+ (ha)	
	Number	Average size	Number	Average size
Arroio Grande	5	12,422	0	
Herval	7	12,582	1	80,000
São Gabriel	40	13,969	5	39,748
Alegrete	65	13,135	3	45,012
Uruguaiana	34	13,362	5	43,124

SOURCE: Data drawn from Chasteen, "Background to Civil War," p. 746.

Rio Grande do Sul partly to its strategic inflexibility in an area continually at risk of invasion by the Spanish.[12] Despite a vibrant wheat trade early in the nineteenth century, by the 1820's agriculture was already in decline. Outbreaks of rust (ferrugem) hastened this, but just as significant were the economic and cultural factors favoring ranching, flourishing in this period with the growing salt-beef trade.[13] In addition, the 1820's introduced the new element of important American competition in the Brazilian flour market, where a superior product milled using modern technology undercut the Riograndense production.[14] Under these circumstances, cultivation produced crops largely for local or regional markets, often in combination with ranching. While in the 1820's there appear to have been subregional differences in the propensity for farming within Rio Grande do Sul, ranching predominated.[15]

By mid-century, the near-exclusive reliance on ranching as a form of land use was well entrenched in the Campanha counties.[16] The council at Alegrete (the largest ranching county in the province) told the ministry of agriculture in 1862 that there was still very little land devoted to the cultivation of crops in their district. Maize, wheat, manioc, and beans were planted for immediate consumption, but very few of the growers had any surplus for sale.[17] Six years later, the councillors informed the ministry of agriculture in a further and detailed exposition that cultivation remained neglected despite the fertility of the land. As an explanation for what they termed the "decadence of cultivation," the councillors pointed to

> indolence, not only of the nationals but also of the greater part of the foreigners (especially the Orientals [Uruguayans] and Argentineans), who abound in the frontier and its neighboring areas. Add to this the special and exclusive taste for the breeding of cattle and you will have the essential causes of the extraordinary lack of development of cultivation. On many breeding ranches there is nothing resembling agricultural work to be found. On these ranches, all of the food products belonging to the vegetable kingdom are bought, when they could be obtained from their own lands with great economy.[18]

This pattern was not exclusive to Alegrete. Around the same time, the members of the council at São Gabriel provided a similar summary view for Rio de Janeiro; although ranching was "the main and almost exclusive industry" of the region, it had not undergone any improvement, so that the system being followed was "still that of primitive times."[19]

There were farmers at work in the Campanha around mid-century, but the human environment was clearly not propitious for them. In 1867, 39 farmers from Dom Pedrito, in the heart of the Campanha, demonstrated how the scales of justice tipped in favor of the larger landowners. Calling for a law to protect agriculture, they complained that they could not "draw the least profit from their daily labor on account of the great quantity of cattle and other harmful animals that wander unbounded, destroying the fences, clusters of trees, and crops."[20] These complaints echoed those of the small farmers of Estreito in the Litoral, who had railed eloquently in 1823 against the unjust distribution of lands and the "laws of iron" their occupiers "imposed on poor agricultural families."[21] Despite complaints made by neglected farmers, the stipulations of the municipal codes continued to reflect the demands of the leading users of land, politically powerful ranchers.

At mid-century, the single most dramatic change in the geography of land use in Rio Grande do Sul was the foundation of German colonies on the forested lands of the Serra. Interested in the prospects for colonization throughout southern Brazil, the imperial government questioned the county councils about available land. In some of the largest Campanha counties, the councillors revealed that they knew their districts enclosed land suitable for cultivation; however, they argued that such areas had little to gain from agricultural colonization as long as they were considered more desirable for ranching.[22]

In the absence of other constraints (and these were many, ranging from political disturbances to diseases), the carrying capacities of the various microenvironments limited the quantity of livestock on the ranches. Based on the observations of the estancieiros themselves, during the early 1840's the general stocking level of the grasslands was placed at 1,500 animals per square league, where cattle were concerned.[23] Cattle predominated on almost all of the ranches.

The heavy emphasis on cattle in the Campanha emerges clearly in a description of the county of São Gabriel, which contained 185 true breeding ranches in 1859, stocked with 508,000 cattle (an average of 2,746 per ranch), 100,000 horses, 100,000 sheep, 1,000 donkeys, and 2,500 mules.[24] This broad pattern still held in 1870. Horses in the area were in a state of visible decline, on account of insufficient attention to breeding. On some ranches the numbers of horses were insufficient to meet even the daily management needs. Sheep were still unimportant, and the lands of São Gabriel were not considered especially suitable for them.[25]

The succession of Major Antonio Guterres Alexandrino of Alegrete, who left an adopted son, Victor, to share his estate with his widow, Ana Joaquina Flora, provides detailed evidence of the types of livestock carried on a single ranch.[26] These inheritors received a ranch with an area estimated at 2.5 square leagues (10,890 hectares). Somewhat less than the area of a sesmaria grant but not yet a fractioned estate, this was a middle-sized property for the time. There were animals of various types present on the ranch, but the bulk of the value of the livestock (69 percent) consisted of 1,114 feral cattle. The property contained some 450 sheep, and it also bred horses and mules.

Most ranches resembled the Guterres Alexandrino estate in carrying a diverse group of livestock beyond a core of feral cattle. Inventories from the period list a wide range of specialized types, such as "burros hechores" (an antiquated and onomatopoeic Spanish term for the donkeys used to impregnate mares in mule breeding) and "cavallos redomões" (horses not yet broken). Despite the long lists of specialized animal types, the bulk of activity revolved around cattle.

Regional variety in the pastoral economy was to a large extent conditioned by the suitability of the natural vegetation for pasturing. At midcentury, the northern part of Rio Grande do Sul was far less developed than the Campanha. Grassland zones of the Serra held few advantages. The accidented terrain provided natural boundaries for many of the estâncias. Travelers found the varied landscape attractive. However, the pastures were of poorer quality and supported only half the number of animals as an equivalent area of the Campanha. As a result, at mid-century the grasslands of the Serra were less dynamic than those of the Campanha; they were still used largely as pastures for droves of mares for breeding mules, and for wintering mules bought from beyond the region. The Cruz Alta region of the Serra alone sold more than 50,000 mules to São Paulo each year.[27] Rio Grande do Sul's administrative geography reflected its ranching ecology; as late as 1860, the entire northern half of the province was divided into only four counties.

Pasture Management, Working Systems, and Technology

Writing during the later eighteenth century, Roscio, who had gained considerable direct experience of southern Brazil through his work as a military official, had tersely summarized the principles of the ranching system being practiced as "that which nature allows."[28] The comment still held much of its force almost a century later. Pastures in Rio Grande do Sul were natural; animals suffered, sometimes to the point of death, when the carrying capacity of the range was exceeded.[29] In the virgin pastures, the grass grew tall enough to conceal the beasts but this situation was becoming less common by mid-century.[30] Overgrazing could lead to the degeneration of the grasslands, especially through the growth of an unpalatable pasture known as "barba de bode" (*Sporobolus sprengelli*).[31] By 1842, alfalfa and low pasture grasses, particularly "capim-de-Angola" and "capim-da-colônia" (both of the family *Panicum maximum*), had already been planted on farms around the towns and larger settlements.[32] These planted grasses showed greater resistance to the winter and were used as forage for the horses and the more valuable cows. But the Visconde de São Leopoldo maintained that the use of planted pastures was generally unknown. There is no evidence of planted pasture use on the ranches themselves. Livestock grazing on pastures deficient in trace elements required salt at intervals in order to fatten satisfactorily.[33]

With very little intervention to improve them, the condition of the natural pastures was the leading constraint shaping the estância calendar. Natural pasturing in turn reflected the seasons. Twin peaks of activity following the extremes of heat in summer and cold in winter marked the work-calendar.

The year opened on an estância with high summer. Attacked by plagues of flies during January and February, the months of maximum summer heat, the cattle declined in health. Ranchers attempted to treat their illnesses, probably using herbal remedies.[34] By March, the insects became less of a problem as the weather started to cool toward autumn. Ranchers began to mark their cattle. Small animals received an identifying sign (*sinal*), a cut in one or both of their ears.[35] Large animals were branded with a red-hot iron, usually on a hindquarter or leg. Ranchers designed their own brands. Dreys described these hieroglyphs in 1839 as "true Chinese bookkeeping" and claimed there were gaúchos who could recognize the ownership of any animal from a glance at its brand.[36] Brands were registered with the town councils. In the absence of fences, the marking and identifying of animals were tasks of key significance. Marking the young stock, which could take weeks on the larger ranches, often provided the excuse for a family festival.[37] The cooler weather of autumn was also the time to castrate the young bullocks, so that they would fatten more easily. There were still some hot, dry days during autumn, favorites for burning the pastures, although some ranchers preferred to leave this for the late winter or early spring.[38] Pastures were burned mainly for the regeneration and control of vegetation, providing what Alexandre Baguet, a Belgian traveler, described as "a sinisterly imposing sight."[39] Burning had further practical purposes. It killed insects harbored by the grasses and it removed the habitats of jaguars (*onças*) and other wild animals prone to preying upon the herds.[40] After burning, new growth came quickly to the range with the cooling rains of autumn, so that the cattle started to fatten up. It was then that the ranches began to organize droves of fat beasts (*manadas*) ready to dispatch for slaughter in the *charqueadas* (slaughtering plants).[41]

The winter months of July and August usually brought gloom to the Campanha, especially when the cold, dry Minuano wind was tearing across the grasslands from the southwest. Lacking forage and buffeted by the winds, the animals would seek shelter in any stands of trees that were available. Their favored variety was the *umbu*, the "grave vegetable sovereign" of the Campanha, as one author has termed it.[42] This grew to enormous dimensions on the grasslands and provided excellent shade. When in season, its fruit also provided important nutrition for the animals; as a result, the tree was deliberately planted around the margins of the small corrals (*potreiros*) where the work animals were kept.

Tender grasses emerged with the spring. At first these purged the animals, and only by the late spring did livestock fully recover from the ordeal of winter. November saw the second peak in the state of fatness of the cattle. Late spring formed a second season when droves could be readied for the cattle merchants.

Fig. 2.1. A cattle hunt in Rio Grande do Sul, c. 1852. From Wendroth, *Album de aquarelas e desenhos*, fig. 70.

The basic method of control on the open range was the *rodeio*, which gathered herds at a specific place for review. Favored spots were always well away from any woods, and there was often a *pau fincado* (a common toponym in Rio Grande do Sul), a wooden post fixed vertically into the ground, to serve as a marker.[43]

Rodeios had a wide range of purposes. The most general was to accustom animals to human presence, fostering their domestication. Sometimes rodeios called only for simple reviews of the condition of the animals, to identify the injured, particularly at the times of intense insect activity. Another task that fell into the straightforward category was the separation of groups of animals destined for slaughter in the Litoral. But other jobs, such as castrating or branding, called for more labor, so that the ranchers found it convenient to help each other. The Visconde de São Leopoldo has left a particularly graphic description of a review rodeio. At dawn, the *capataz* (manager) would ride out with a sufficient number of peons, four or six, according to the work to be done. Boys (Indians or salaried slaves from eight to twelve years in age) were often used; their limited weight did not tire the horses to the same extent as that of grown men. These workers went out onto the range whistling and shouting; packs of trained dogs helped in gaining the attention of the animals. At these signals, the already domesticated cattle would respond by running with "marvelous instinct" to the post, where the tasks of the rodeio were performed.[44] Wild cattle posed a greater challenge.

As much or as little as individual ranchers required, the peons herded these to foster their domestication. There was a strong social aspect to the rodeios. Earlier in the century, Luccock had noted that "in a country so little enlivened by variety, this assemblage [formed] one of its most rural and pleasant scenes."[45]

The frequency and quality of rodeios were not constant. In the early nineteenth century, the tax-contractor Manoel Antônio de Magalhães had complained that even wealthy ranchers economized on operating expenses by leaving most of their cattle "as wild as the bulls of Portugal."[46] By mid-century, rodeios were far more regular events. The Conde de Piratini informed the manager of his Estância da Música in 1832 that he wanted the cattle worked in the rodeio ("costeado") as frequently as possible. His caprice showed in the command that the labor should sleep near the places of the rodeios in the summer, so that they could commence their work of treating the animals against the ravages of insects at the break of day.[47] Based on what ranchers had told him, the Visconde de São Leopoldo claimed in 1842 that rodeios took place once a week.[48] By 1869, Manoel Pereira da Silva Ubatuba, Rio Grande do Sul's veterinary inspector, was complaining to the president of the province about the habit of repeating rodeios unnecessarily. Animals driven "by the hooves of the horse and the teeth of dogs," in a manner that was "more an entertainment than a task of work, more an act of perversity [*judiaria*] than a cure-treatment," could only demonstrate a weakened resistance to the winters. Ubatuba claimed they received even more brutal treatment in branding. The ranchers could see the poor state of the cattle but placed routine on a higher plane, so that before the beasts were branded, they were chased, lassoed, flung to the ground, and dragged along. After the cattle were burned with the red-hot iron, the whole process was often repeated to give pleasure to those who had not yet had the opportunity to engage in the chase or to use the lasso. In Ubatuba's view, those following careful practices were more the subject of gossip than of imitation.[49]

On the open range, animals were bound to stray at intervals into neighboring properties, especially at times of drought. On such occasions, groups of peons (*recrutadores*) were sent to fetch them back. This required asking a neighbor for a rodeio in order to find the missing animals. The councils were very specific about the circumstances under which ranchers might deny the right of rodeio to their neighbors. Those who did not follow the county rules carefully with regard to such matters as branding, reviewing, and making up *tropas* of fat animals for delivery to the charqueadas ran the risk of being fined or spending a spell in the county jail.[50]

At mid-century, the technology used in ranching was still that of the cattle hunts of colonial times. The gaúchos accomplished their work almost exclusively on horseback with a limited selection of equipment, chiefly the lasso (*laço*), the balls (*bolas* or *boleadeiras*), and the knife. All of these made handy weapons as well as forming the tools of the ranch. The lasso and balls likely entered Rio Grande do Sul with the Guaraní Indians, who used them to hunt game, especially rheas.[51] There was considerable address displayed in the manufacture and use of these tools. For example, the lasso, a braided cord with a noose (both made from raw leather), needed to be flexible in order to be effective. Tallow grease rendered it too soft. Either a piece of fresh meat, especially the liver, or manure extracted from the bowels of a cow while still warm was considered the ideal lubricant. The other major implement, the bolas, consisted of three balls (often, but not exclusively, of stone) covered in leather. Two of the three balls were of the same size, linked by a cord (*soga*) a meter to a meter and a half long. In the middle of this cord, the third and smaller ball (the *manicla* or *manica*) was attached with a soga of around half the length of the other.[52] Seizing the manicla in the right hand (the same hand used for the lasso) the gaúcho would set up a rotating motion for the two larger balls, building sufficient momentum to hurl and entwine them around the neck, back, or hindquarters of an animal. It was not difficult to damage an animal if the bolas were flung with less than complete skill. Practicing their use from early childhood onward, however, the gaúchos excelled in the use of these implements.[53] Nicolau Dreys claimed that they could thus master any kind of animal, including the jaguars of the grasslands.[54]

Ranch Labor

In the early nineteenth century, the label "gaúcho" generally fell on those who lived on the range beyond authority and fixed settlement, or, as a common saying had it, without faith, king, or law. By the 1850's the term applied more specifically to the peons who performed the work on the estâncias. Given the nature of ranch work and in a world where everybody rode, the social distances between the various segments of rural society were undoubtedly much narrower than in Brazil's plantation zones. Nevertheless, the Campanha had a social hierarchy, at whose apex stood the estancieiros. They were not a homogeneous group.

TABLE 2.2

Labor on the Ranches of Selected Counties of Rio Grande do Sul, c. 1860

County	Number of ranches	Capatazes	Free peons	Slaves
Rio Pardo	40	32	34	173
Alegrete	391	124	159	527
São Borja	568	171	339	153
Jaguarão[a]	239	107	—	243

SOURCE: "Mappa numerico das estancias existentes nos differentes municipios da Provincia, de que até agora se tem conhecimento official, com declaração dos animaes que possuem, e crião por anno, e do numero de pessoas empregadas no seu costeio," *maço* 532, AHRGS.

[a]At Jaguarão some, although not all, free peons were included in the number of slaves. There are no separate statistics on the number of free peons.

While many were still running their ranches directly at mid-century, the wealthiest were often absentees. Absentees required reliable managers (capatazes). Beneath these capatazes came the peons. Within the peon hierarchy the *domadores*, those capable of the exacting work of "breaking" horses and mules, stood in relief. For the rest, the leading division was between the core of permanent staff on a ranch and the floating population of casual labor drawn in to perform specific tasks.

At mid-century, the administration of many ranches still fell under the direct control of members of the landowning families. Where capatazes were called upon, they may have managed more than one ranch (Table 2.2). It was difficult to retain good capatazes; during the frequent instances of political insecurity along Brazil's southern frontiers, many of them were quick to appear in active service in the National Guard. In some instances, the managers had considerable freedom in the running of the ranches. In general, however, the lines of their authority were drawn much more narrowly. The Conde de Piratini on his Estância da Música, for example, instructed his manager that any debts were to be left until his occasional visits. Cesar has aptly observed that this control system, which left all the economic initiative in the hands of an occasional visitor, stifled the economic development of the Campanha.[55] Landowners who visited their properties infrequently preferred their capatazes to be able to write, and at least some could.[56] The salaries of the capatazes were linked with their degrees of responsibility, reflected most obviously in the sizes of the herds under their care. For example, accounts for the Barão de Cambaí's ranches in 1870 show the manager at Cambahy (3,420 breeding cattle) receiving 24$000 a month and his counterpart at Santa Eugênia (11,606 cattle) 64$000. However, all of this patriarch's managers were in debt to him for food and goods received, so that the recording of their salaries became largely an accounting exercise.[57] Few ranchers paid wages on a regular basis.

Capatazes sought out peons with the skills to work as domadores; this was recognized as a particularly arduous and dangerous life. According to the Visconde de São Leopoldo, domadores frequently suffered accidents from horses and mules that rebelled against their training; such animals remained useless for riding or breeding.[58] Whatever their background, domadores commanded respect.

The peons on Riograndense ranches had gaúcho origins; as such, all of the authorities agree that they had a mixed heritage. Many peons were mestizos; others were Indians who had sought work on the ranches after the destruction of the Jesuit missions. African slavery, a key institution for Brazil as a whole, has usually been considered of limited significance on the ranches of Rio Grande do Sul. However, scholars in various parts of the Americas are beginning to question the long-held tenet that slavery and ranching were generally incompatible. Modern research has revealed that recourse to slave labor on the ranches of Buenos Aires Province at times of labor shortage was stronger than previously assumed.[59] In Rio Grande do Sul, despite Cardoso's excellent work exploding the myth that there was a racial democracy among the gaúchos along the southern frontier, the state of knowledge about slavery in the region remains very incomplete, especially for the Campanha. References to the trade in slaves between the southern frontier and other parts of Brazil are almost always couched in vague terms. It is clear, however, that slaves were being moved southwards in some number during the early decades of the nineteenth century as a response to labor shortages.[60] By 1833, there were at least five thousand African slaves in Pelotas, which led some to call the place the "black Euphrates."[61]

The question of how many such slaves ended up on the ranches has been less satisfactorily resolved. Within Rio Grande do Sul it is usually claimed that slave labor was of only minor importance on the ranches.[62] The basis for this claim lies largely in the contradictory evidence gleaned from travel accounts.[63] On the other hand, Spencer Leitman has argued convincingly that ranch slavery was important for at least the first half of the nineteenth century.[64] Archival evidence gathered for this study confirms that slaves were important for the operation of ranches at mid-century and even beyond.

In 1850, the imperial government ordered the military commanders along the southern frontier to survey the condition of the Brazilian-owned ranches lying within the territory of northern Uruguay. The 36 estâncias to the south of Chuí at the tip of Lagoa Mirim contained over 200 slaves.[65] Around 1860, the government of Rio Grande do Sul produced a survey of the ranches in the province. This reveals that slaves outnumbered free peons on the ranches in some counties (Table 2.2), including Alegrete, whose grasslands carried the largest cattle herds in the entire province.[66] Most counties sent incomplete answers to Porto Alegre; the four shown in Table 2.2 provided fuller records than any others. Even in these counties, the staff levels per ranch seem very low indeed (see below). It is probable that respondents counted only their cores of permanent staff; there is no indication in the manuscript whether the survey included casual workers, such as agregados and boundary peons (*posteiros*). In addition, marginal comment against the entry from Livramento (not shown in Table 2.2 on account of its incompleteness) noted that many capatazes and peons were not included since they were away on duty in the National Guard, a reflection of Brazil's unstable political position with Paraguay at the time. The low numbers of staff per ranch in the counties shown in Table 2.2 make it safe to assume that this political instability was felt more broadly than in Livramento alone. Thus the levels of free labor around 1860 were probably lower than the norm. However, if free labor was called away to serve in military exercises, this can only have increased the degree of dependence on slaves. A recent conclusion that slaves were "a fortuitous factor of production" on the ranches is misleading.[67] Slaves formed a significant part of the labor force on the ranches.

That said, dependence on slave labor did vary, even within a single district. On the Brazilian-owned ranches in the Chuí section of the Uruguayan frontier with the Campanha, the largest slave populations in 1850 (twenty, seventeen, and sixteen) were to be found on large properties (with areas of seven, thirteen, and five square leagues, respectively). Yet the largest estância in that region, which encompassed some 49 square leagues and held a supposed cattle population of 100,000 head, had only fifteen slaves present.[68] Since the military survey described only 36 properties, it would be wrong to attach too much significance to the numbers of slaves in relation to the sizes of the estâncias or the numbers of stock carried. The dominant impression left by the Chuí survey is that no general relationship can be established between the sizes of the ranches and their complements of slaves. By no means all of the slaves were clustered on giant properties. For example, in 1849 Benigno José de Souza of Bagé left his descendants a ranch of around 3.5 square leagues, home to 54 slaves (39 men and 15 women), of whom 10 were *roceiros* (cultivators) and 14 *campeiros* (cowboys).[69]

Some modern Brazilian authors, such as Guilhermino Cesar and Décio Freitas, have accepted the presence of slaves as an important element on the estâncias but still see them as having been involved in the auxiliary operations, such as growing food, rather than the key activities that demanded skilled riding ability.[70] Their argument is based more on conjecture than on evidence. Freitas, for example, has posited two major obstacles to the use of slave labor in the central pastoral operations; these are the lack of technical experience in the arts of ranching on the part of Africans and the problem of supervision along an open frontier. Freitas's argument that "nothing could be more anti-economic than the importation of a black from Africa in order to submit him to a long training" will not stand.[71] Boys born of slave mothers in Rio Grande do Sul during the earlier decades of the century were grown men by 1850; there had been plenty of time for them to witness and to learn pastoral techniques. And at mid-century, slavery was still a firmly entrenched institution in Brazil.

The roles customarily ascribed to the slaves along the southern frontier cannot be limited with such certainty. A listing forwarded to Brazil's foreign minister in 1850 of slaves who had escaped out of Rio Grande do Sul includes no shortage of entries under the label of "campeiro." Details about individuals provide clues beyond this generic term, revealing that slaves were much more directly involved in ranch work than is usually assumed. Alexandre, for example, a mulatto, had the characteristic of spread toes (the gaúchos placed only the big toe in the stirrup) from the habit of continual riding, the truest single mark of the gaúcho. Manoel's owner described him as dexterous with horses and able to work on the range without problems. Further gaúcho characteristics appeared in Julio, said to be "very dexterous with the lasso, much given to drinking and to gambling."[72] Inventories also provide details that sometimes lend clues to the status and occupations of the slaves. For example, on the ranch of João Silveira Gularte (Goulart) in Bagé, his slave João, a Creole aged 21, was assessed in 1849 as suitable for "all the work of the range and horse-breaking."[73]

The numbers of slaves on the ranches provide a crude but useful indicator of the effect of regional variety in ranching on capital accumulation. In 1850, there was an average of 5.5 slaves on each of the Brazilian-owned ranches within the Chuí frontier of northern Uruguay.[74]

Around 1860, within Rio Grande do Sul itself, the numbers ranged from 4 slaves per ranch at Rio Pardo to as low as 0.3 per ranch at São Borja in Missões (based on the data in Table 2.2). Some of this variation can be explained. Rio Pardo was associated with established wealth; as a garrison town, it had been closely involved with the early military cycle of settlement in the Campanha. Missões, by contrast, was still isolated at mid-century, poorly situated for the transport of its cattle to coastal slaughter; ranchers there used Indian labor instead of African slaves.[75] Capital accumulation was slower and slaves were generally less numerous in the Serra than in the Campanha.[76]

Slaves were a presence on some of the ranches long after 1850; sources drawn from these later decades help to clarify the connections between slavery and the "traditional" ranch. In 1876, an Uruguayan who visited the huge ranch of Francisco Pereira de Macedo, the Barão (and later Visconde) do Serro Formoso, found large-scale cultivation of cereals there, on a "primitive system." He noted that there were 45 to 50 slaves present to perform the labor but added the qualification, "as much for the pastoral tasks as for the cultivation."[77] José Maria da Gama Lobo Coelho d'Eça, the Barão de Saican, registered 57 slaves at São Gabriel during 1872 in connection with the Law of Free Birth; no fewer than 23 were specifically described as cowboys (see Table 2.3).[78] Further down the gradient of rancher wealth, the proportion of the slaves given over to ranch labor was probably lower. Of the eleven slaves left by Francisco de Assis Brasil (father of the famous rural "improver" Joaquim Francisco) in 1872, three were described as campeiros in their registration.[79]

A comparison of the 23 cowboys with the 14 other male slaves ascribed occupations in Table 2.3 shows that they were younger (with an average age of 34 as opposed to 56) and that a far higher proportion of them had been born in Rio Grande do Sul (87 as opposed to 29 percent). Moreover, the data contained in Table 2.3 are detailed enough that they indicate the presence of three generations of slaves belonging to the Barão de Saican in 1872. No fewer than eighteen of his slaves (or 43 percent of those born within Rio Grande do Sul) can be traced from his single freed slave Eufrasia (see Table 2.4). As soon as they were old enough to work, all of Eufrasia's male descendants became cowboys (Tables 2.3, 2.4). Three generations of slaves represents a settled pattern of ranch labor, one that takes on features of a community. At least in the Barão de Saican's case, the established family ties of the slaves go far to circumvent Freitas's objection that the problem of supervision would have hindered ranch slavery.

TABLE 2.3

Slaves Belonging to the Barão de Saican in 1872

	Color	Age	Marital state	Place of birth	Parentage	Aptitude for work	Occupation
In São Gabriel, mostly on the Estância de Santa Maria							
Felicianno	Black	82	Single	Africa	Unknown	Little	Kitchen gardener
Antonio	Mulatto	69	Widowed	Bahia	Unknown	Little	Carpenter
Domingos	Black	62	Single	RGS	Unknown	Regular	Field laborer
Amaro	Black	61	Single	Africa	Unknown	Sufficient	Cowboy
Miguel	Black	61	Single	Africa	Unknown	Sufficient	Cowboy
Ciriaco	Black	60	Single	Africa	Unknown	Little	Field laborer
Sebastião	Black	60	Single	Africa	Unknown	Regular	Field laborer
Francisco	Black	57	Single	Africa	Unknown	Sufficient	Cowboy
Alexandre	Black	53	Single	Africa	Unknown	Sufficient	Field laborer
Izabel	Black	51	Married	Africa	Unknown	Regular	Laundress
Marcos	Black	39	Single	RGS	Son of Izabel	Sufficient	Cowboy
Manoel	Black	30	Single	RGS	Son of Izabel	Sufficient	Cowboy
Guiximbe							
Maria	Black	50	Single	Africa	Unknown	Regular	Seamstress
Simplicio*ᵃ*	Black	36	Single	RGS	Son of Maria	Sufficient	Cowboy
Justina	Cabra*ᵇ*	48	Single	RGS	Daughter of *Eufrasia*	Regular	Cook
Bernardina	Cabra	26	Single	RGS	Daughter of Justina	Regular	Laundress
Julia	Black	8		RGS	Daughter of Bernardina	None	None
Antonio	Mulatto	7		RGS	Son of Bernardina	Little	None
Manoel Vieira	Mulatto	23	Single	RGS	Son of Justina	Sufficient	Cowboy
Porfirio	Black	20	Single	RGS	Son of Justina	Sufficient	Cowboy
Eufrasia	Mulatto	20	Single	RGS	Daughter of Justina	Regular	Seamstress
Magdalena	Mulatto	13	Single	RGS	Daughter of Justina	Little	Seamstress
Esperança	Mulatto	12	Single	RGS	Daughter of Justina	Little	Seamstress
Cirino	Cabra	10		RGS	Son of Justina	Little	None
Belarmino	Mulatto	40	Single	RGS	Son of *Dominga*	Little	Shoe-maker
Eugenia	Mulatto	37	Married	RGS	Daughter of *Eufrasia*	Regular	Baker
Felippa	Black	18	Single	RGS	Daughter of Eugenia	Little	Laundress
Vicente	Black	2		RGS	Son of Felippa	None	None
Marianna	Black	17	Single	RGS	Daughter of Eugenia	Little	Seamstress

Name	Color	Age	Marital status	Origin	Parentage	Condition	Occupation
Asipio	Mulatto	10		RGS	Son of Eugenia	Little	Cowboy
João	Mulatto	8		RGS	Son of Eugenia	Little	None
Jannario	Mulatto	34	Single	RGS	Son of *Felizarda*	Sufficient	Cowboy
Manoel	Mulatto	32	Single	RGS	Son of *Eufrasia*	Sufficient	Cowboy
Pacifico	Mulatto	31	Married	RGS	Son of *Felizarda*	Sufficient	Coachman
Lino	Black	30	Single	RGS	Son of Guiseria (dead)	Sufficient	Cowboy
Maximiano	Mulatto	30	Single	RGS	Son of *Dominga*	Sufficient	Cowboy
Jorge	Mulatto	28	Single	RGS	Son of *Dominga*	Sufficient	Cowboy
Liberato	Mulatto	28	Married	RGS	Son of *Rita*	Sufficient	Cowboy
Amancio	Mulatto	23	Single	RGS	Son of *Rita*	Sufficient	Cowboy
Fernanda	Mulatto	18		RGS	Daughter of *Luzia*	Regular	Starcher and ironer
Magdalena	Mulatto	4		RGS	Daughter of Fernanda	None	None
Luisa	Mulatto	15	Single	RGS	Daughter of *Afra*	Regular	Servant
Candido	Mulatto	13	Single	RGS	Son of *Afra*	Regular	Cowboy

On the Fazenda Pirajú in Itaqui

Name	Color	Age	Marital status	Origin	Parentage	Condition	Occupation
Antonio	Black	60	Single	Africa	Unknown	Little	Butler
Manoel	Black	60	Married	RGS	Unknown	Regular	Cowboy
Matheus	Black	54	Single	RGS	Unknown	Sufficient	Cowboy
Joaquim	Black	50	Single	Africa	Unknown	Little	Field laborer
João Burro	Black	50	Single	Africa	Unknown	Little	Field laborer
Custodio	Mulatto	42	Single	RGS	Son of *Felizarda*	Sufficient	Cowboy
Thomáz	Mulatto	36	Single	RGS	Son of *Thomazia*	Little (sick)	Cowboy
Eleodoro	Mulatto	22	Single	RGS	Son of *Thomazia*	Sufficient	Cowboy
Ricarda	Mulatto	20	Single	RGS	Daughter of *Thomazia*	Sufficient	Seamstress
Reinalda	Mulatto	18	Single	RGS	Daughter of *Thomazia*	Sufficient	Seamstress
Marcos	Cabra	16	Single	RGS	Son of Justina	Sufficient	Cowboy

On the farm in Santa Maria

Name	Color	Age	Marital status	Origin	Parentage	Condition	Occupation
Felippe	Black	60	Single	Africa	Unknown	Little	Field laborer
Adão	Black	58	Single	RGS	Unknown	Regular	Field laborer
Domingos	Black	50	Single	Africa	Unknown	Little	Field laborer

SOURCE: Inventory of deceased: José Maria da Gama Lobo Lobo Coelho d'Eça, Barão de Saican; executrix: Baronesa de Saican; São Gabriel, 1873, N 9, M 1, E 108, APRGS.

NOTE: Names of slave mothers who did not form part of the Barão de Saican's property in 1872 are italicized.

[a]Slave escaped.

[b]A light-skinned person, the child of mulatto and black parents.

TABLE 2.4

Slaves Belonging to the Barão de Saican Who Were Descended from Eufrasia (Freed) and Their Ages in 1872

Children of Eufrasia	Grandchildren of Eufrasia	Great-grandchildren of Eufrasia
Justina (48)	Bernardina (26)	Julia (8)
	Manoel Vieira (23)[a]	Antonio (7)
	Porfirio (20)	
	Eufrasia (20)	
	Marcos (16)	
	Magdalena (13)	
	Esperança (12)	
	Cirino (10)	
Eugenia (37)	Felippa (18)	Vicente (2)
	Marianna (17)	
	Asipio (10)	
	João (8)	
Manoel (32)		

SOURCE: Inventory of the Barão de Saican, São Gabriel, 1873, APRGS.
[a]Names of cowboys appear in italics.

Most of the contemporary sources indicate that labor needs were low on the ranches during the first half of the nineteenth century. The assertion that four to six peons could work a ranch of three square leagues is common.[80] Leaving the cattle feral was the leading solution for those who sought labor economy. In 1842, the Visconde de São Leopoldo maintained that he had heard many times from "good practical ranchers" that the same number of workers needed for 2,000 cattle could usually manage nearly 3,000.[81] However, labor needs in some of the rodeios may have been much larger; one authority on local customs maintained that they needed between 20 and 30 men.[82]

The wage-bill of an individual property illustrates the cycle of labor needs and costs more concretely. A team of a capataz, a core of three salaried peons, casual staff, and possibly some of his thirteen slaves worked Major Antonio Guterres Alexandrino's ranch. In the period from 1 February 1856 to 15 October 1857, the cost for the regular wage labor on the ranch (the capataz and three peons) summed to over a conto, charged against the production costs. The burden of fixed salaries appears quite high (Table 2.5). For the seasonal tasks of castrating and marking, the cast of characters widens considerably, with work being performed by Antonio Gonçalves, his son-in-law Feliciano, and the

TABLE 2.5

Labor Costs on the Ranch of Major Antonio
Guterres Alexandrino (Alegrete), 1856–57

Regular labor

José Joaquim Lucas, capataz, 20.5 months at 32$000	656$000
Israel Antonio Lucas, peon, 20.5 months at 12$000	246$000
Manoel Syrino, peon, 5 months at 12$000	60$000
Antonio Candido da Rosa, peon, 4 months at 8$000	42$000
TOTAL	1:004$000

Casual labor

Antonio Gonçalves, Feliciano (his son-in-law), and Athanazio for help with marking animals during 1856	16$000
Antonio Gonçalves, Feliciano, and Athanazio for help with castration during 1856	6$000
Zacarias, Manoel, Orfilio, Athanazio, and Rufino for help with marking during 1857	14$000
Athanazio, Porfirio, Rufino, and Americo de Mello for help with castration during 1857	8$000
TOTAL	44$000

SOURCE: Inventory of deceased: Antonio Guterres Alexandrino; executrix: Ana Joaquina Flora; Alegrete, 1853, N 117, M 8, E 65, APRGS.

other workers Athanazio, Zacarias, Manoel, Orfilio, Rufino, Porfirio, and Americo de Mello. Despite the breadth of this rather poetic list of names and the fact that some of these men were hired more than once, the cost of casual labor amounted to only 44$000 in the same period of a little more than twenty months running from 1 February 1856.

There are signs of a probable relationship between the supply of slave labor and the quality of ranch management. While the roles of the thirteen slaves in the service of the Guterres Alexandrino family are impossible to determine from the wage-bill alone, the availability of slaves may explain in part why the cost of casual paid labor was so low on the Guterres Alexandrino ranch. On Benigno José de Souza's Bagé ranch all of the 1,450 cattle present at mid-century were domesticated rather than feral, with 350 of them explicitly described as confined to the corral.[83] This degree of management was exceptional; however, his 3.5-square-league estância had a huge staff of 54 slaves, 14 of whom were cowboys.

Estância Production

Estâncias mainly worked to produce steers for slaughter in the char-queadas. The traditional age for killing a steer was not less than four to five years or more than seven. Younger animals lacked the greater degree of profitable fat and the larger hides of the more mature animals. In addition, it was important that the meat of the animals have acquired a sufficient degree of consistency to withstand a strong salting. Cows were killed as well, whenever they were unable to bear more calves or had difficulty entering into pregnancy.[84]

Even without considering management questions, the physical diversity of Rio Grande do Sul had a profound impact on the proportion of the cattle the ranchers could afford to market each year without it affecting the size of their herds. Saint-Hilaire, questioning and critical, was excellent on this point, receiving very different testimonies around 1820.[85] His conclusion that Campanha ranchers could expect to sell one-tenth of their herds is remarkably close to that provided by the councillors at São Gabriel, who made a rare effort to quantify the local production during 1859 (Table 2.6). Since the councillors reckoned that 508,000 cattle stocked the ranches of their county, the sale of 45,000 indicates that the ranchers sold an average of around 9 percent of their herds in that year. The council estimated that the value of the exports from São Gabriel summed to 1,337:000$000 (roughly 140,000 pounds sterling by the values of the day). Cattle and their products accounted for no less than 88 percent of this sum.

In the accounts for a single Alegrete estância covering most of 1856–57, the rate of production was more modest still.[86] The core (*casco*) of the herd belonging to the Guterres Alexandrino family consisted of 494 cattle in early 1856. After 129 calves were marked during 1856 and 133 in 1857, the core of the herd had grown to 756 cattle by late 1857. The 264 mares on the ranch had given 68 "crias" in 1856 (these could be either foals or mules, presumably) and 69 in 1857, resulting in a horse population of 401 animals. Sheep, almost always the poor relations of the regional animal kingdom, gave a less impressive performance. Mainly on account of *peste* afflicting the lambs, they "suffered delay in production." Between the sheep and their surviving lambs, the size of the flock was reckoned at around 330 animals in October 1857.

In this period of nearly two years, the ranch sold only 39 steers, 8 percent of the original core of the cattle herd. These steers were split from

TABLE 2.6

Exports from São Gabriel, 1859

Product	Unit value	Total value
Beef-cattle (40,000)	24$000	960:000$000
Cows (5,000)	20$000	100:000$000
Mules (2,000)	16$000	32:000$000
Horsehair		25:000$000
Hides		120:000$000
Timber and boards		100:000$000

SOURCE: CM of São Gabriel, doc. 312a, 1859, AHRGS.

the main herd to form a drove (tropa) and sold to Zeferino Vieira for 1:170$000 (or 30$000 each), which would have been enough to meet the wage labor costs of the estância. In mid-October 1857, after some steers were sold and others fed to the slaves, the Guterres Alexandrino herd stood at 667 head, including both the tame and feral cattle ("entre mansas e chucras"), as well as two oxen.

The Guterres Alexandrino family may have deliberately sold a low proportion of the cattle in 1856–57 in an effort to rebuild their herd. Major Antonio Guterres Alexandrino had owned 1,265 cattle in 1853; the numbers declined after his death with the sale of animals, probably to meet legal expenses. No doubt Campanha ranchers expected to sell more in years of favorable pasturing conditions, but the evidence reviewed here reveals that 10 percent was a realistic target for cattle around mid-century.

While the character of the range set the basic parameters of production, there were other constraints. Three of these—disease, the breakdown of management practices, and robbery—were particularly important. Cattle disease lay beyond the control of the traditional rancher at mid-century; poor management and robbery were linked to patterns of human behavior at times of political instability.

References to *epizootia* (contagious disease) reveal that this became a serious yet sporadic problem in the early 1840's.[87] It had been so bad in 1843 that soldiers and travelers could not camp by muddy water holes due to the smell of putrefying animal corpses. Over the years, disease reappeared intermittently, especially in the grasslands close to the Serra. Cattle disease was regionally selective. It appears to have particularly affected animals on the move, especially the cattle brought great distances from Uruguay to restock Riograndense ranches after the Farroupilha

Revolt. By the later 1850's, it had occasioned sufficient damage that the provincial president sought information from ranchers and other interested parties about its symptoms.

The answers reveal that some of the ranchers had major problems with the cattle tick (*carrapato*).[88] They had little sense of what caused this, or what to do about it. Since the ranchers generally knew nothing of veterinary science, they ascribed all deaths to the peste. According to General Andréa (provincial president of Rio Grande do Sul, 1848–50), large numbers of diseased animals, including breeding cattle, were sold off to the charqueadas.[89] Carrapatos gave the ranchers something else to complain about, but they managed to live with the problem rather than subjecting it to scientific scrutiny.

Provincial officials were quick to criticize the ranchers for what they viewed as poor management techniques, especially the temptation to kill prematurely whenever prices were buoyant. There were many in Rio Grande do Sul open to changing the regular pattern in the ranching calendar; in 1842, the Visconde de São Leopoldo attributed this to "the ambitious, who are killing whether or not the cattle are fat."[90] Some of the ranchers did not wait until their cattle had reached even three years of age before they sent them for slaughter. At mid-century, a period of recovery from war, horses were a focus for specific official concern. Horses were slaughtered mainly for their hides, although Andréa lamented that thousands of mares were killed each year for their grease alone; the grease served to make soap or as a fuel for street lighting. However, the evidence coming from the Campanha councils shows that patterns of horse slaughter were not as indiscriminate as Andréa implied.[91]

However much in the public interest, council efforts to regulate ranch production were clearly difficult to enforce, not least because the estâncias could be located tens of leagues from the county seats.[92] Poor ranching practices were largely a reflection of disturbed political conditions, compounded by the incidence of cattle diseases. As provincial president João Lins Vieira Cansansão de Sinimbu observed in 1854, whether the ranchers were interested in regaining their social position or in paying off their debts, the means to achieve their ends were always the same. Any animals remaining in a given locality were shipped off to the charqueadas without much scruple about age or ownership.[93] In turn, poor ranching practices led to breed degeneration, which carried the long-term implication of depressed production.[94]

A third major constraint on production was robbery, in large part a reflection of rural poverty.[95] Some of the wholesale destruction of Brazilian property in the interior of Uruguay during the Guerra Grande had important consequences for the Campanha. The persecution of the Brazilian ranches to the north of the Río Negro by General Manuel Oribe's frontier commanders during the later 1840's (and the seizure of their animals to pay political debts in Argentina) had led to a steady drift of refugees into the Campanha.[96] Around 1850, animal theft was taking place on a large scale. In an effort to reduce theft, the government of Rio Grande do Sul congregated men without family, occupation, or estância work in São Gabriel or Porto Alegre, supporting them there from public funds.[97]

This public initiative to control social disruption cannot have been totally successful. In 1853, a rancher from Alegrete apportioned much of the blame for management difficulties on the ranches to the rural unemployed. While the peste, war, and a major drought in 1840 had combined to reduce the number of breeding animals in the province, nothing had depressed production so strongly as robbery, which claimed entire herds. The Alegrete rancher concluded his analysis by lamenting the weakness of government authority in the interior: "Those who have no cattle (the greater number today) understand that those who do should support them. This is the law which is being executed, notwithstanding that it is not written, while our legal codes sleep, or are insufficient against this *communismo*, so comfortable for the vagrants and so ominous for the upright ranchers."[98]

The councillors at Bagé were sufficiently concerned about cattle robberies in 1857 that they asked the president of the province to revise the penal code. Complaining that the spate of thefts had seriously damaged the rights of private property holders, the council wanted the theft of animals of all kinds to be subject to the same kinds of penalties as other forms of robbery, implying that animal thefts were previously viewed more leniently.[99] From an analysis of food costs made by the councillors of Alegrete in 1857, it is clear that large-scale cattle theft (*abigeato*) went hand in hand with the rising price of meat, a reflection of the strong markets for salt-beef during the 1850's. As meat became a more valuable commodity, ranchers ceased giving it freely to the poor. Nor were ranchers at Alegrete any longer prepared to offer the poor land on their estâncias on which they could set themselves up and grow crops.[100] Ranch life was starting to change.

Ranch Profits and Finances

Most of the data on the profits and finances of nineteenth-century ranches in southern South America are highly speculative. In 1855, a Chilean historian reckoned the annual return on capital on the pampas of Argentina to be as high as 30 percent, but that simply equated profitability with the rate of reproduction of the herds.[101] Leaving aside the awkward matter of return on fixed capital, a cruder derivation of profitability based on the balance of annual receipts over expenditure, drawn from highly limited cases, has led Jonathan Brown to conclude that Argentinean ranches could yield 20 percent profitability rates in good years.[102] Even this method of estimating profitability has numerous difficulties; the facts that ranchers often slotted extraneous items of domestic consumption into estância expenditure, or held back animals from the market to build up herds, will serve as but two examples. It is unlikely that many Campanha ranches yielded profitability rates as high as 20 percent.

Based on production, operating costs, and an estimate of the value of the property, in 1821 Saint-Hilaire had reckoned the return on capital for a ranch in the western Campanha (one that used wage labor) at 8 percent.[103] Around 1850, contemporary observers identified a division of interests in the pastoral economy of Rio Grande do Sul.[104] Those involved in processing and marketing ranch products had incomes which far outstripped those of most of the ranchers, many of whom lived in permanent debt. *Charqueadores* held the great bulk of the liquid capital in the province, and they used the resulting power to hold the ranchers in dependence.[105] While ranchers numbered in the thousands, there were only several dozen charqueadores providing the vital outlets for ranch products. Without a specialized organization to press for their group interests, individual ranchers established their own working relationships with the charqueadores. Charqueadores advanced credit to ranchers, who in turn were obligated to send their animals to specific slaughtering plants. Some of these arrangements probably worked well, especially within extended family networks, not least by clarifying the outlets for ranch production. But as Antônio Manoel Correa da Câmara observed in the 1840's, ranch production was a type of gambling, where luck rarely held beyond a few games. The previous section reviewed some of these "gambles," such as drought, the peste, and frontier political instability. Correa da Câmara laid particular emphasis on the consequences of partible inheritance; with subdivision, most estates passed along the generations in what he termed a "state of mediocrity and dependency" very close to penury.[106]

There were ranchers in southern South America who were spectacularly successful; the Anchorenas of Argentina are the leading example. While their estates had been developed using commercial capital in the early nineteenth century, they came to depend exclusively on ranch profits for the generation of further investment funds. As Brown has carefully stressed in his study of the formation of the Anchorena cattle empire, this was hardly the typical ranching family; it was unique in terms of scale.[107] However, the Campanha probably shared some of the characteristics of the Anchorena empire.

As in Argentina, it is likely that the most successful Campanha ranchers at mid-century were those who came to the business with management skills and with access to development capital. The Conde de Piratini, for example, had owned a charqueada as well as estâncias.[108] Besides being a rancher, the Barão de Cambaí was a successful private banker. Like the Barão de Mauá, a more famous entrepreneurial son of the south, it is probable that he learned this craft in a Rio de Janeiro commercial house, where his parents had sent him to work as a young man.[109] By his death in 1869, the Barão de Cambaí had advanced letters of credit summing to almost 83 contos, and the chain of his debtors extended through much of the Campanha and into Uruguay. The Cambaí estate had almost as much cash on hand.[110]

Within the Campanha, capital available for estância development was concentrated in few hands. The best evidence for this comes from São Gabriel, where a short-lived savings bank was founded in 1853 with the objective of doing away with the usury that existed in the district.[111] Some years later, the councillors from the same region lamented that they could not contract a loan for public works because capital was very scarce ("muito raro") in the Campanha.[112] In ranching, as in other economic spheres, established riches were often a sure route to further wealth.

Material Culture and Patterns of Life in the Campanha

At mid-century, the ranch houses had evolved beyond the leather huts described by authors of the colonial period, but most were still extremely simple structures. The results of slow improvement in the quality of houses are evident in inventory descriptions. While the Carvalho family of Alegrete described one of their multiple properties in 1848 as an "old estância of palm," where everything was "much ruíned through its length of service" (a common phrase in Riograndense inventories), they owned other estância houses with stone walls and tile roofs, surrounded by planted orange and lemon trees.[113] Orchards were common on Campanha ranches. The tiling of roofs and plastering of walls did not alter the general character of estância houses; even in the 1880's, an English traveler could still describe them as

> not much more than barns, generally built of mud and bamboo, and roofed with what is known on English lawns as pampas grass; they are entirely without anything approaching the ornamental, either inside or out. They are, of course, always one-storeyed, and contain one living room, in which are a table and a row of chairs. Beyond are bedrooms, somewhat better furnished, the bed linen edged with home-made lace. I have seen an English advertisement of lager-beer nailed to the wall as a picture in one house; but as a rule floors and walls are of mud, and if the latter were even whitewashed, it is rarely done a second time.[114]

There were a few exceptions to this general simplicity. Robert Avé-Lallemant, a Lübeck doctor, registered his evident pleasure in 1858 at seeing the "large white house with two lateral staircases" that belonged to the Barão de Cambaí.[115] Not too long after mid-century, the Visconde do Serro Formoso owned an estância house with Italianate decoration in the Campanha.[116] But the most powerful landowners, those with diversified economic interests, often made their principal residences somewhere other than their estâncias.[117] By 1832, the Conde de Piratini, for example, was already living in his houses at Pelotas and at the port of Rio Grande, visiting his rural property only periodically in order to adjust the accounts.[118] The Barão de Saican had overseen his more than 90,000 hectares of land in the Campanha from his 26-room *sobrado* (a house with more than one story) in the main square ("praça da matriz") of São Gabriel.[119]

Most ranches contained little of interest by way of material goods; equestrian objects were frequently the most valuable things to be found. For example, among the things valued in 1849 on a Bagé ranch were silver spurs and a pair of stirrups made of "old plate with the weight of two pounds."[120] The stark simplicity of the material culture of most of the people living on the estâncias left strong impressions on visitors to the region. Travelers often slept on hides, for example. Yet all visitors stressed the high degree of hospitality generally shown; when laying the foundations for an estância house, it was customary to trace the lodging for visitors as the first portion.[121]

The broad pattern of food consumption on the part of ranch labor did not vary much.[122] Although many ranchers encouraged (and sometimes insisted on) the small-scale cultivation of crops, meat, particularly beef, was by far the most important food on the ranches. Roasted either on spits or in holes in the ground (as the *churrasco* or *assado*), it was consumed in large quantity. The councillors at São Gabriel claimed in 1859 that meeting the food needs of that parish consumed 15,625 cattle and 10,000 sheep each year, impressive quantities when the population had been placed at just under 5,000 the year before.[123] On Major Antonio Guterres Alexandrino's ranch, the thirteen slaves belonging to the property had consumed 10 cows and 40 bullocks in twenty or so months.[124] In addition to any sheep, each Campanha dweller was likely eating at least two to three cattle per year. Samuel Gottfried Kerst reckoned the assado had an effect on sociability similar to that of the appearance of champagne on European tables.[125]

A second invariable element in the diet was an infusion of *erva mate* or Paraguayan tea (*Ilex paraguariensis*), brought down from the Serra, where it grew in profusion. By habit, the peons daily sucked up several pints of this from a gourd through a pipette (*bomba*), usually made of silver. *Mate* also had the practical benefit of giving peons the resilience to cover long distances on horseback without stopping to eat. Greetings over, *mate* was always the initial thing offered to the caller at a ranch; Avé-Lallemant described it rather well as "the symbol of peace, of concord, and of complete understanding."[126]

The Serra provided more than *mate* at mid-century; for example, Alegrete also drew part of its supply of maize, beans, and flour from there, "with great difficulties in the transport."[127] Despite the poor transport throughout the province, the diet of the landowners (and perhaps of some of their staff) was broadening. In 1870, the accounts presented by the Alegrete merchants Leite & Silva for the Barão de Cambaí's large Santa Eugênia ranch include such items as beer, English butter, conserved fish (including shrimp) and jams, hyson tea, and the local cane alcohol (*aguardente da terra*).[128] Where slaves were present on the estâncias, they were sometimes the objects of particular dietary attention on the part of their owners. For example, both the Conde de Piratini and the Barão de Cambaí catered to their penchant for tobacco.[129]

Dress was distinctive enough to attract plenty of comment, but since it was in evolution its exact character at mid-century is hard to fix.[130]

Fig. 2.2. A gaúcho on horseback, c. 1870. Photograph by Marc Ferrez. Reproduced from Hoffenberg, *Nineteenth-Century South America in Photographs*, p. 117.

Gaúchos always wore broad-brimmed hats, made of felt, straw, or palm-leaf. In winter, woollen ponchos in vivid blue or red, often striped, formed universal garb to fend off the Minuano. The *chiripá*, a type of girdle made of coarse, domestic fabric, wound between the legs, fastening to a belt at the front. Sometime in the nineteenth century, the characteristic *bombachas*, baggy trousers probably without par for equestrian comfort, replaced it. Again, while at the beginning of the nineteenth century most of the gaúchos were still riding barefoot, using spurs fixed to their heels by strips of raw hide, by mid-century tall leather boots were becoming customary. Visitors to the Campanha never failed to comment on the size of the spurs (*chilenas*), with their giant rowels that rendered walking difficult and noisy. A red neckerchief often made a fitting accompaniment to the colorful ensemble of the gaúcho.

For the few who showed interest in a broader world, isolation marked life in the Campanha.[131] Even wealthy estancieiros showed lack of knowledge not only of major political happenings but also of basic local geography; on his trip through Rio Grande do Sul of 1845–46, Baguet claimed that one businessman in the western Campanha had asked him if he would have to pass through Mexico in order to return to Brazil![132] Church played a limited role. In 1856, Luís Alves de Oliveira Belo, a politician, described the church at Itaqui as "indecent" on account of its poverty.[133] The liberal elites foresaw a civilizing influence in the establishment of churches; thus the councillors of Bagé called in 1847 for the creation of a chapel at Dom Pedrito to subtract a section of the Brazilian community in northern Uruguay "from the dangerous contagion of immorality" fostered by war.[134] Most of the rural population probably never entered a church, alive or dead. A large fig tree close to an estância house often marked the site of a simple family cemetery.

The towns were small, but with more ranchers moving off their estâncias, at mid-century they were growing rapidly in the western Campanha. In 1856, Itaqui had 40 houses covered in tile and 50 to 60 others roofed in straw. Eight ranchers were seeking to build there in the near future.[135] Despite the limited populations of towns like Alegrete and São Gabriel, their commercial movement impressed visitors by its intensity. Commercial establishments in the Campanha towns were well stocked, so that the ranchers had little need for extensive movement. For example, in the tiny place of Carajasinho, which was little more than a *venda* (store) and an estância headquarters, the store contained everything the local population would require, "from the Parisian shoe and silk parasol to the colossal iron spur of the peon."[136] Such establishments must clearly have catered to the entire social hierarchy. In addition, the ranchers were served at their doors by what the council at Bagé termed in 1853 an "inundation" (*aluvião*) of peddlers, bringing contraband at lower prices from Montevideo on their pack-mules or in ox-carts.[137] Goods brought to the frontier of the Campanha from Uruguay were 20 to 30 percent cheaper than when imported through the port of Rio Grande.[138]

There was already a veneer of social activity in the towns that seemed familiar to European visitors. In São Gabriel, Avé-Lallemant received numerous invitations to attend piano recitals in private homes; in addition, the ambience of the monthly balls held in Uruguaiana struck him as "perfectly European." This German doctor claimed he could see boundaries in the landscape between the zones affected by "universal European civilization" and those reflecting different values, such as Missões. Based on what Avé-Lallemant discerned in 1858, the reach of "Europeanization" extending from the Litoral had already affected São Gabriel more than Alegrete in the western Campanha.[139] During the remainder of the nineteenth century, that "Europeanization" would begin to permeate not merely the Campanha towns but also the estâncias.

Standing at the crest of a coxilha, a mid-century traveler looked down on a kaleidoscope of animal activity. Numerous cattle, horses, and rheas filled the vista as far as the horizon, but the human imprint remained barely visible on the landscape. While some ranchers showed very clear concern for quality in their production—it is unlikely that traditional Riograndense ranching was as backward as is sometimes indicated—access to land on a large scale was still the key determinant of the system.[140] The physical potential of the range did not equal that of the humid parts of Buenos Aires or southern Uruguay, but at its best it had given scope for some to grow rich, more often in cattle and slaves than in any material comforts. Even in the mid-century traditional ranching economy, values were far from static. But the following six or seven decades were ones of pivotal challenge, coming especially from the coast.

CHAPTER THREE

The Coastal Complex

IF THE ESTÂNCIA was the leading ranching institution for the Campanha, its counterpart for the Litoral was the charqueada. This had a shorter history than the ranch and its organization was in flux in the early part of the century, as the markets for ranch production broadened. In southern South America, industrial technology first appeared on a significant scale in the slaughtering plants; as they were the satellite points of contact between the metropolis and the periphery, their history is particularly important in any effort to understand the region's development.[1]

When the main markets for Riograndense hides and meat lay far beyond the region, in the absence of sophisticated land transport such as railways, large-scale slaughtering was economically feasible only in locations close to the Atlantic, or along rivers that allowed ready access to the coast. Animals destined for slaughter held the advantage that they could be walked on the hoof to the sites of the charqueadas. The imported salt used for meat preservation was another major factor explaining the coastal locations of the charqueadas; on account of the costs of salt's transport, Pedro de Alcântara Bellegarde, a soldier-diplomat, concluded in 1849 that it was "impossible to *charquear* (make salt-beef) in places distant from the ports."[2] Only limited amounts of beef were dried in the wind on the estâncias themselves (the so-called *charque-de-vento*, which used little salt); this dried beef supplemented the supply of fresh meat. In addition, the hides stripped from animals killed in the interior for domestic consumption were moved to the coast by *carretas*, high-wheeled ox-carts.

During the early nineteenth century, most of Rio Grande do Sul's charqueadas were concentrated very close to the coast, along the plain of the São Gonçalo and its tributaries (see Figure 3.1). By the 1820's, a distinctive culture emerged around the town of Pelotas. Here, in contrast to the Río de la Plata, the plants were established and controlled

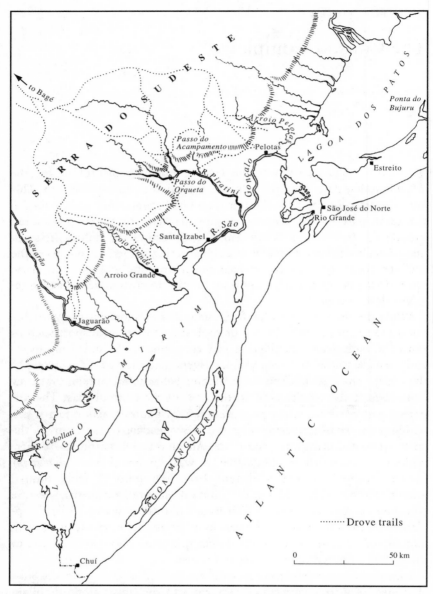

Fig. 3.1. The Litoral region of Rio Grande do Sul. Partly after Lopes and Azevedo, orgs., *Carta geographica do Estado do Rio Grande do Sul.*

almost exclusively by Luso-Brazilians, settlers who quickly built family dynasties (sometimes termed the "tallow aristocracy" in imperial Brazil) and displayed a very strong attachment to their immediate region. Some slaughtering plants, however, had been established further north along the Jacuí Valley near Rio Pardo (Figure 1.1). These Jacuí Valley char- queadas offered shorter cattle drives to the ranchers of northern Rio Grande do Sul, but they were more than four hundred kilometers re- moved from the Atlantic through the Lagoa dos Patos. The establish- ment of charqueadas in such a distant location reflected the fact that during the first half of the nineteenth century Rio Grande do Sul's salt- beef industry had confronted weaker competition from the Río de la Plata than it faced in 1850, in part because of the great political insta- bility associated with the independence struggle around the Plata. At mid-century, the location constraints in Rio Grande do Sul were tight- ening as competition from the Plata increased. Unable to absorb prof- itably the transport costs to Rio Grande, the point of export, the Jacuí Valley charqueadas were falling into disuse during the 1850's, leaving those of the Pelotas area predominant in the industrial geography of Rio Grande do Sul. In 1854, there were 25 charqueadas working at Pelotas.[3] Many of the other industrial establishments there, such as tanneries, bone-ash works, and a mare-oil refinery (where animal grease was turned into fuel for street lighting), clearly reflected the importance of the cattle economy.[4]

Given the predominantly coastal pattern of slaughter, many cattle were subjected to lengthy and expensive drives. Arriving tired and thin, after journeys that could last for more than twenty days, the animals yielded less in their carcasses. Since any mismanagement of a drive (tropa) ate easily into the profit margins of a ranch, many estancieiros supervised the transport of their animals directly. On their way toward the coast, the drove-trails generally avoided rivers by following the spines of the coxilhas. However, in order to reach the Litoral from the Campanha, animals needed to cross the Piratini, a major river in the higher ground of the Serra do Sudeste (the leading passes are shown in Figure 3.1). Councillors at Pelotas complained in 1854 that the difficul- ties of transporting the ox-carts and droves across the Piratini formed a major obstacle to economic development.[5]

Animals for sale were congregated at a place known as the Tablada, part of the plain of the São Gonçalo about two kilometers from the town of Pelotas, where their weight and value were assessed by eye.[6] Under normal circumstances the killing season (safra) in Rio Grande do

Sul extended from 15 December to the middle of June. However, weather and abnormal market conditions influenced this calendar. During the safra, cattle fairs took place six days a week at the Tablada, making this the most important point of congregation for those involved with ranching in the province. As many as 500,000 cattle were killed in a year, although the harvest did not usually run beyond 400,000 beasts. Charqueadores maintained their flow of production by purchasing a significant number of their animals from northern Uruguay; the proportion of the slaughter supplied from there fluctuated according to the availability of cattle within the Campanha.

The Work of the Charqueadas

The best early description of a Riograndense charqueada was provided by Luccock, who described the industry in its formative stage, shortly before the foundation of Pelotas in 1811:

> When the cattle are killed and skinned, the flesh is taken off from the sides in one broad piece, . . . sprinkled with salt, and dried in the sun. In that state it is the common-food of the peasantry in the hotter parts of Brazil . . . and as it will keep long forms an excellent sea stock, and would bear carriage to distant parts of the world. Some idea of the immense quantity of beef thus prepared may be formed from the fact that, in one year, an individual, José Antonio dos Anjos, slaughtered fifty-four thousand head of cattle, and charqued the flesh.[7]

More technical descriptions of the work of making salt-beef in Rio Grande do Sul are rare; most travelers kept away from the plants, concentrating on the repugnance they felt both for the actual operations and for the treatment of their labor force of slaves.[8]

Nicolau Dreys provided a convincing account of the nature of the work in 1839. At that time some of the charqueadores were still using archaic methods to kill cattle, which ran along the following lines. Two peons would work in conjunction. One approach had a peon aggravating an animal to exhaustion (mainly by waving a red poncho in its face) within an open corral; another method depended on the lasso. The role of the second peon was to cut the articulation of the legs (using a lance with a blade shaped like a half-moon), either at the point of exhaustion or at the moment when the beast attempted to free itself of the lasso. In both cases, the animal was left to die on the ground. Even by 1839, some processors had opted for a greater degree of system. Dreys summarized their motives as to economize on labor, reduce the risks of the

Fig. 3.2. A Pelotas charqueada in 1825. Jean Baptiste Debret provides a rare glimpse of the organization of a charqueada. Note the considerable area devoted to the drying of hides and salt-beef. Note also the varying levels of technology present. For example, while the charqueada uses a crane to move carcasses from the slaughtering floor (right foreground), the grease plant, from which smoke billows, still has no chimney. From Debret, *Viagem pitoresca e histórica ao Brasil*, fig. 89.

work, and minimize "the repugnancies inseparable from the act and consequences of death."[9]

The improved system reflected the beginnings of mechanization. Cattle were lassoed to a moving post (a type of windlass moved by human power) and dragged along a corridor, at the terminus of which the slaughterer maintained a fixed position. As soon as its head pressed against the boundary fence, the animal was knifed with an iron stiletto at the beginnings of the first collarbones, bringing immediate asphyxiation. A revolving crane immediately moved the dead animal onto the *cancha*, a tiled surface; usually slightly convex in shape, these facilitated the drainage of the blood. On the cancha the hide was stripped and the animal was bled and cut apart. Dreys recorded that the charqueadores attempted to screen the sordid aspects of the slaughter under a veil of greenery for which they used the leaves of the *butiá*, a type of coconut palm (*Cocos jatahy*).

In another part of the establishment (the *salgadeiro*), the flesh of the carcass was salted in large pieces (*mantas*). The salt came from a variety of sources, including northeast Brazil, Portugal, Cape Verde, and various other Atlantic islands.[10] After a strong salting, the meat was stacked in piles with intermediate layers of salt. Compressing the meat in large stacks accomplished the first phase of its drying; in addition, excess salt drained in solution from the stacks under the pressure of their weight.

This salt-solution, stored in a reservoir beneath the stacks, was used for preserving secondary products, such as ribs and tongues. The final stage in turning the salted beef into charque was the completion of the drying; this took place on a large expanse of land covered with rows of wooden frames, where the mantas were suspended to dry through the action of the sun and the winds. Once completely dry, the charque was cut into large, oblong pieces and stored on a platform as it awaited shipment, protected from the weather by a covering of hides. The time taken to produce charque clearly depended on the weather; under favorable drying conditions, the whole process required a little over a week.

Hides stripped from the animals were cleaned of any remaining bits of meat or fat and then either stacked on the ground to dry or placed in a tank containing a salt-solution drawn off from the meat during preservation. From the tank, the hides were stacked, like the meat, with layers of salt between them. The degree of care taken in the drying and folding of hides probably reflected their value in the markets. In the *graxeira*, often a separate part of a charqueada, heads and bones were boiled up in large cauldrons, in order to extract grease, which was frequently stored within the stomachs or large intestines.

Around mid-century, a steer destined for slaughter normally weighed around eight *arrobas* (120 kilograms), with the hide considered its leading product.[11] Even within the same region, the sizes of hides were subject to considerable variation, and data on their customary weights are very sparse. Those stripped from steers were the largest; the Visconde de São Leopoldo maintained in 1842 that 28 pounds (12.7 kilograms) was their usual weight in Rio Grande do Sul.[12] Using the limited British consular price data available, in 1855 the value of a hide from a typical Riograndense steer ranged from a low of 3$360 for a salted hide sold in January to a high of 9$660 for a sun-dried hide sold toward the end of the year; most of the hides had sold toward the higher end of the price range.[13] In addition to its hide, a steer would usually provide around 60 kilograms of charque. Based on the average sale price of charque for 1855 (4$000 per arroba), each steer yielded salt-beef worth some 16$000.[14] The markets were not always this buoyant; by the early 1860's, the average sale price of an arroba had declined to 2–3$000. Still, it is clear that charque was much more than a fortuitous by-product accompanying the hide; it counted as a primary product.

Most of the salt-beef was sold without being packaged in any way, but some was packed into barrels. Jonathan Brown has implied that packing meat into barrels was customary in Buenos Aires, but the extent of this practice in Rio Grande do Sul remains obscure.[15] In the statistics Antônio José Gonçalves Chaves compiled on Rio Grande do Sul's exports from 1805 to 1820, irregular (but small) quantities of charque packed into barrels appear throughout the period, almost all prepared around Porto Alegre.[16] The markets for salt-beef in barrels were invariably Rio de Janeiro and Montevideo. The small proportion of the meat packaged in this way points to a highly specialized use; charque in barrels may have been used for military supplies, perhaps as food on board ships. The Visconde de São Leopoldo maintained in 1842 that charque packed into barrels rather than sold in piles was more lucrative for the processors since they could make use of the bones. What he meant by this is unclear; certainly the grease extracted from bones could be packed into barrels as a distinct by-product. At mid-century, the packaging of charque in barrels was taking place, but still on a minor scale.

Each steer yielded seven to eight kilograms of tallow and grease; these were the most important secondary products, and they found broad consumption. Tallow was used for making candles and soap, while animal fat in the form of grease was in universal use for cooking. The potential for profit did not end here, however. A charqueador could sell the salted ribs, as well as the so-called "smalls" (*miúdos*), such as the heart, liver, salted intestines (tripe), and hooves (turned into *mocotó* oil), as food for slaves.[17] Other products found profitable sale for specialized artisan uses in Europe. Hair, for example, was used for stuffing sofas in Victorian homes, or by plasterers for mixing in with their mortar.[18] The British sought bones and bone ash as the first of the "artificial" fertilizers; by raising crop yields on the lighter soils of Lincolnshire and eastern Scotland, South American bones played a key role in the nineteenth-century agricultural revolution.[19]

Charqueada Infrastructure and Improvements

Although they were finding a commercial use for almost every part of the animals that they slaughtered, the salt-beef plants were still generally very unsophisticated places at mid-century. Buildings and other fixed installations were usually simple. Charqueadas required land for their operations and for the congregation of animals awaiting slaughter. They needed a location on navigable water for shipping out their products. The importance of river access was reflected in their form. For example, Zeferina Gonçalves da Cunha's Pelotas charqueada fronted on

TABLE 3.1

Capital Investment in the Charqueada Graça, 1853

Investment	Mil-réis
Land of the Fazenda da Graça	60:000$000
House, charqueada, and other buildings	15:000$000
Slaves (81)	59:950$000
Yachts (2)	6:500$000
TOTAL	141:450$000

SOURCE: Inventory of deceased: João Simões Lopes; executor: João Simões Lopes Junior; Pelotas, 1853, N 366, M 26, E 25, APRGS.

the São Gonçalo for 305 meters but ran back from the river for 1,760 meters; it adjoined other plants on each side of its river frontage.[20]

Mário José Maestri Filho has recently described the "insufficiency of the known sources" about the level of capital investment in the charqueadas as "profound."[21] Nevertheless, inventories help to clarify the mix of investment. The data contained in Table 3.1 are drawn from the estate of Commander João Simões Lopes, a Portuguese who had founded the Charqueada Graça around 1800 and died shortly after midcentury. Table 3.1 overstates the investment in the salt-beef plant in several ways. Graça was also a working ranch carrying several thousand animals. The charqueada establishment valued at fifteen contos also included the ranch buildings, likely of little value. Some of the slaves may have been working on the ranch or in Simões Lopes's domestic service, but it is likely that most of them were busy in the charqueada. Still, Graça represented a considerable investment, more than six times greater, for example, than that in one of the more important sugar plantations of Rio Claro, São Paulo, around the same period.[22] Slaves represented a very high proportion of the investment in the charqueadas. Based on the 1868 valuation of a plant at Jaguarão, Maestri Filho has estimated the "buried" investment in slaves at 55 percent of a total of approximately 40 contos, but this property, sold in connection with a mortgage, may not have been in a flourishing condition before the sale.[23] The investment in slaves was a higher proportion of the total at the Charqueada Graça, probably at least 70 percent. Slaves represented a far larger part of the capital investment than any equipment.

Thus apart from land and ready access to navigable water, the other key ingredient of a salt-beef plant was the brute strength of a labor force made up almost exclusively of slaves. Domingos José de Almeida testified in 1843 at Alegrete that he had owned 84 slaves before the Far-roupilha Revolt, of whom 72 worked in his charqueada.[24] We have seen that João Simões Lopes left 81 slaves in 1853. The picture of extensive reliance on slave labor grows even clearer in the inventory of Antônio José Gonçalves Chaves made in 1872; no less than 48 of the 53 slaves recorded in the inventory were working in the charqueada, 16 as slaughterers and the remainder as general laborers.[25]

Contemporary observers, foreign and local, entertained no doubts about the profitability of the charqueadas. The Visconde de São Leo-poldo noted in 1842 that charqueadores could expect to recoup their investment within six to eight years, depending on how industrious they were.[26] Facing ranchers short of working capital, often a result of the property divisions that followed from partible inheritance, charquea-dores sometimes took over land directly.[27] A strong link developed be-tween Pelotas salt-beef processing and landownership; that link ex-tended beyond the Campanha into Uruguay.[28] Apart from Graça in Pe-lotas, the only ranch Commander João Simões Lopes owned was a large estancia of eight Spanish leagues in Uruguay.[29]

Inventories support the contemporary impression that charqueadores were wealthier than most of the ranchers. For example, the land at Graça with its plant installations and slaves constituted only 29 percent of João Simões Lopes's net wealth at his death; he left an estate worth 478 contos in total (approximately 57,000 pounds sterling by the values of the time). The Simões Lopes estate included more in money and shares (nearly 145 contos) than in the investment in Graça. This pattern is repeated to an even more striking degree in the inventory of Antônio • José Gonçalves Chaves (a second-generation salt-beef processor) (Table 3.2). Had the Gonçalves Chaves partnership chosen to place its financial assets in another form of investment, they would have sufficed at the time to buy one of Rio de Janeiro's large coffee plantations with all its assets.[30]

Although Mawe had asserted during the early nineteenth century that Rio Grande do Sul furnished charque superior to that drawn from the Río de la Plata, by 1850 nobody claimed this any longer.[31] While the salt-beef plants throughout southern South America remained generally primitive affairs at this time, some improvements to systemize produc-tion had begun around the Plata. Most of the key innovations were made by Antoine Cambaceres, a French immigrant chemist who arrived in Buenos Aires in 1829. Cambaceres introduced assembly-line tech-niques, such as the use of cattle chutes in the corrals and truck and rail

TABLE 3.2

Assets of Antônio José Gonçalves Chaves and
João Maria Chaves, 1871

Asset	Value in mil-réis
Charqueada (land, buildings, and other fixtures)	40:000$000
Slaves (48)	42:160$000
Yacht	5:000$000
Total value of charqueada assets	87:160$000
Money on hand	156:721$382
Shares (mainly in municipal bonds and local navigation companies)	54:500$000
Letters of credit	50:096$482
Other monies lent (including bank deposits)	174:526$443
Life insurance policies	4:501$306
Total value of financial assets	440:345$613
Balance of assets (including a ranch, livestock, and three houses)	253:039$700
TOTAL value of the estate	780:545$313

SOURCE: Inventory of deceased: Antônio José Gonçalves Chaves; executor: João Maria Chaves; Pelotas, 1872, N754, M45, E25, APRGS.

systems for the movement of carcasses around the *saladeros* (slaughtering plants). He also designed steam vats, which extracted grease from bones and flesh more effectively in terms of both quantity and speed than mere boiling.[32] These organizational and technical changes led the Pelotense charqueadores to refer to a "Platine system" of producing salt-beef.

Researchers on the Plata have argued that some of Cambaceres's improvements gained rapid acceptance throughout the region and that the use of steam was generalized by the later 1840's.[33] In Rio Grande do Sul, a periphery of the southern South American ranching region, the patterns of adoption were uneven. Although an 1839 description of the Pelotas charqueadas certainly implies that some had already introduced a greater degree of system in their establishments, improved techniques of production had been adopted only partially by mid-century in Rio Grande do Sul.[34] The correspondence of the charqueador Domingos José de Almeida makes this clear. Examining a plant near Porto Alegre in 1836, Almeida described the property as still killing animals with the lasso in a large corral, with the clear implication that this was an old-fashioned practice.[35] Almeida himself, using steam at Pelotas before

1835, represented a much more advanced technical stage.[36] But he was an exceptional figure in both ability and circumstance. In the 1820's he speculated heavily, bringing slaves and goods in his own vessels from northeast Brazil to Pelotas, no doubt in the expectation that Brazil would win its struggle to retain the Cisplatine Province (Uruguay). He lost heavily in these transactions and by the early 1830's was pressed by his creditors to squeeze a higher return from his charqueada, hence his adoption of steam.[37] The potential for further technical improvement at Pelotas was undermined in 1835 by the outbreak of the Farroupilha.[38] During a decade-long civil war that saw shifting patterns of territorial control, the rationale for investment in fixed equipment was weak. It is likely that improved techniques were only beginning to find generalized use at mid-century.[39]

A Difficult Passage to Market

Salt-beef plants located on the São Gonçalo around Pelotas had the disadvantage of lacking direct access to the sea. Charqueadores moved their products by launch to the nearby ocean port of Rio Grande, the true point of export. Difficult of access from the Atlantic, Rio Grande was a poor port, where the shifting sands and the shallow water of the coast made navigation treacherous.[40] The increasing size of ships and the silting of the principal entrance of the Rio Grande bar had reached such a point by mid-century that almost all vessels of heavy draft needed the help of a steamer-tug to cross. These ships had to wait for the tug and favorable weather conditions in order to reach the port.

The limitations of the port did not end when a vessel had crossed the bar; there were bureaucratic as well as navigational difficulties at Rio Grande. There were two anchorages, one at Rio Grande itself and a second, able to accommodate larger ships, at São José do Norte. Until 1858, the Brazilian government maintained customhouses at both places, but in that year it closed the house at São José. Consul Henry Vereker described what this meant for the foreign vessels (notably those of the British, who were using larger ships) that were taking on what formed the heavier cargoes for the region:

> Vessels with cargoes consisting wholly of salt or coals, are allowed to continue to discharge all their cargoes in the anchorage of São José do Norte, which is much better than the southern anchorage, but vessels with general cargoes are required to discharge at the south (Rio Grande) being allowed to unload a sufficient portion of their cargo at the São José anchorage to lighten

Fig. 3.3. The port of Rio Grande, c. 1852. From Wendroth, *Album de aquarelas e desenhos*, unnumbered fig.

their draught, so that they may come through the shallow channel to the south, the cargo thus discharged being immediately brought over to Rio Grande in lighters. Bone-ash and ballast are usually completely laden at São José; and thereat, as the deeper anchorage, most British vessels complete their cargoes of hides or other merchandize.[41]

Bureaucratic problems stem from human nature and could be overcome. The problems for trade caused by the sands and shallow waters of the Rio Grande do Sul coast were much more difficult to deal with. By the 1850's, the merchants and ranchers of the region were already keenly aware that the lack of a good port at Rio Grande acted as a brake on the development of the ranching industry.

The Markets for Pastoral Products

Despite the headaches of navigation, the export trade in Rio Grande do Sul's ranch products was generally vigorous at mid-century. Its markets were divided between the artisan trades of the industrializing nations of the North Atlantic and the plantation complex of the Americas.

TABLE 3.3

Exports of Leading Animal Products from Rio Grande do Sul to Europe and the United States, 1855

Article	Quantity	Article	Quantity
Cattle hides (dry)	331,822	Shinbones	559,100
Cattle hides (salted)	210,697	Hoofs, kg	42,342
Horsehides (dry)	4,836	Wool, kg	35,898
Horsehides (salted)	21,021	Hair (mixed), tons	453
Calfskins	21,938	Bones and bone ash,	
Horns	691,271	tons	10,679

SOURCE: Vereker's report on Rio Grande do Sul for 1855, H. of C. 1859, session 2, XXX, p. 404.

The long-distance trade, dominated by hides (Table 3.3), was bound for two principal broad destinations, Europe and the United States, with some shipments delivered through intermediate ports, such as Antwerp or Cádiz. As around the Plata, dry hides, the more valuable commodity per piece, sold mainly in the United States, while salted hides found their principal market in Britain.[42] The basis for this division in the hide market is not clear; since tanning was still very much an art, it may have been linked to differences in the methods employed in these two markets.[43] One authority maintained that the initial impetus for salting hides in Rio Grande do Sul came from Britain; the salting of hides was first tried there as an experiment ordered by the British mercantile firm Holland Davies.[44]

Hides shipped from the Plata fetched higher prices in the markets than those from Rio Grande do Sul for being larger, heavier, and more even in size; this reflected the larger creole cattle to be found there, in turn a result of the superior natural conditions for ranching. However, prices were based on reputation as well as the quality of the raw material.[45] For example, the hide merchants in the United States sought a lighter article than the English, one that Rio Grande do Sul was well suited to supply. Hides from Rio Grande do Sul generally did not match the prices of those from the Plata, nor were they received in the markets on the same footing.[46]

Large quantities of other animal products were also exported (Table 3.3). Rio Grande do Sul joined the Río de la Plata to form Britain's major supplier of horns, used to make handles for tools, umbrellas, and utensils, as well as for the manufacture of combs and buttons.[47] British farmers seem to have been keen to buy all the South American bone ash that became available; from the limited data available on Britain's total bone supply during the 1850's, Rio Grande do Sul alone was responsible for around 10 percent of it.[48] By 1860, continued growth in the demand for this fertilizer meant that the stock of bones in Rio Grande do Sul was diminishing, except for large deposits in the interior left by the early-nineteenth-century charqueadas of the Jacuí Valley, which were still deemed uneconomical to work.[49]

Charque was by far the most important product entering into coastal trade, most of which was conducted with Rio de Janeiro and the cities of northeast Brazil. In 1856, 16,720 tons of charque were shipped in coastal trade from Rio Grande. Although a new market opened up for them in Britain after 1861, grease and tallow, too, were still destined for domestic consumption around mid-century, with 1,154 tons of grease and 835 tons of tallow dispatched from Rio Grande in 1856.[50] The main markets for salt-beef lay with the expanding plantation economies of Brazil itself and of parts of the Caribbean, where it was used principally as food for slaves. Charque from Rio Grande do Sul competed with a similar product drawn from the south, and the destinations for Riograndense salt-beef fluctuated. Thus, around 1820, charque produced in Pelotas was being exported mainly to Rio de Janeiro, Bahia, and Havana. In 1842, the Visconde de São Leopoldo noted that while charque had been taken to Europe, the Cape of Good Hope, and Havana, its principal market still lay with Brazil's major coastal cities.[51] A few years later, Bellegarde claimed that competition from the Plata had pushed Brazil out of the Havana salt-beef market and had made the product cheaper in Rio de Janeiro and Bahia.[52] The loss of the Cuban market was important; at mid-century, that island was in the middle of a 40-year expansion in sugar cultivation extensive enough to be called its Golden Age.[53] Even within Brazil, Riograndense charque met serious competition in the northeast from Newfoundland salt-fish.[54] This Newfoundland trade had developed after the opening of Brazil's ports to international shipping in 1808. With the cost of food on Brazil's plantations increasing steeply during the 1850's (in part as a reflection of diminishing subsistence cultivation), salt-cod was becoming a more important item in slave diets.[55] In 1858, a good year for the trade, New-foundland shipped a little more than 20,000 tons of salt-fish to Brazil, almost all of it to Pernambuco and Bahia.[56] The viability of the Newfoundland salt-fish trade was clearly linked with charque prices, which in turn were a reflection of the quantities of salt-beef coming into the market.[57]

Broad Assessment of the Narrow Economy

Whether ranching began its modernization in the Campanha from a socioeconomic base commensurate with that of similar areas of the Río de la Plata is not easily established, since the evidence is often contradictory. Early in the century, some of the foreign observers who passed through Rio Grande do Sul were optimistic about its economic prospects, basing their judgment in large part on the visibly rapid growth of Pelotas and its organized cattle slaughter.[58] Brazilian commentators tended to be much more pointed in their criticisms. By the 1850's, the general tenor of comment on Rio Grande do Sul was negative, whatever its source.

Early in the nineteenth century, some observers pointed to the benefits of relative political stability, benefits enhanced by the political turmoil in the Río de la Plata of that period.[59] However much the ranchers in the Campanha may have distrusted some policies of the Brazilian government, at least Brazil had the apparatus of a nation-state, and this was to their advantage in the early decades of the nineteenth century.

The best of the local critics were better placed than any foreign travelers to set the local pastoral economy in a broader context. They were well aware of its wealth but could also see its drawbacks. Gonçalves Chaves, the Pelotas charqueador and talented political economist, pointed in the 1820's to the socially inadequate and economically inefficient distribution of lands, the insufficient attention devoted to cultivation, and the limited basis for international competition in the marketing of ranch products; this last he attributed to the mishandling of a peripheral economy by the central powers of Rio de Janeiro. In an economy that was an appendage to the center, where several producing areas were entering into competition for a specific market, there was a need for sensitivity to the fiscal structure of the neighboring zones.[60] By the 1850's, the validity of the prescient calls for reform of the 1820's reemerged.[61]

During the political disruption of the 1840's around the Plata, the Pelotas salt-beef plants had successfully competed with those based in Buenos Aires and Montevideo.[62] Some charqueadores moved their slave labor in search of politically viable temporary locations.[63] However, the greater political stability of the 1850's was generally perceived as unfavorable for Rio Grande do Sul's exports, as the region began to experience renewed commercial competition from Argentina and Uruguay.[64] The return to a generalized peace brought a deeper interest in technology in the labor-short Río de la Plata. In addition, the opening of new slaughtering plants along the Paraná and Uruguay Rivers offered a major source of competition for Rio Grande do Sul.[65] During 1851–55, the quantity of hides exported from Rio Grande fell by nearly a third.[66]

Symptoms of difficulty began to emerge clearly in 1858. The price of hides had fallen by 48 percent from the previous year. Livestock was so "superabundant" that the province was no longer able to consume all it produced, yet the import-export trade of the port was described as daily losing force.[67] In large measure, trade difficulties were reflections of commercial problems in the international economy. For example, the financial crisis that began in the United States at the end of 1857 depressed the profit levels of the planters of coffee in Brazil and of sugar in Cuba, thus limiting the principal markets for salt-beef.

These broad trends were clearly understood by the president of Rio Grande do Sul.[68] Others saw this commercial weakening through a narrower lens. The leaders of the Commercial Association at Rio Grande attributed the economic decline to competition from Uruguay and to the illicit trade practiced between there and the interior of Rio Grande do Sul.

In 1857 a new Treaty of Commerce and Navigation between Brazil and Uruguay freed Uruguayan salt-beef from import taxes in Brazil.[69] This measure further damaged prospects for economic development in Rio Grande do Sul.[70] It made the Uruguayan product more competitive in the Brazilian market, but it also benefited Uruguayan producers of salt-beef by enabling them to attract Campanha cattle sent south for slaughter, then return their products north for consumption in Brazil. Uruguay's intermediate trade (the reexport of products drawn from Argentina and Rio Grande do Sul) was undoubtedly growing stronger during these years. Although farther from the markets for pastoral products, Montevideo was a better natural port for international shipping than either Buenos Aires or Rio Grande. Between 1856 and 1858, half of the hides and a quarter of the salt-beef passing through the port of Montevideo bound for Brazil and Cuba originated in Rio Grande do Sul and Argentina's Litoral.[71]

The continued growth of the herds had further intensified the commercial crisis by 1862. In that year, salt-beef sold for only a third of its value in 1861. One deputy to the provincial assembly estimated that over 100,000 fat cattle on the estâncias of Rio Grande do Sul had gone unsold during the past year. Nonetheless, in the financial year 1861–62, 153,000 more cattle had been killed than in the previous one. Between

1857 and 1862, the number of animals killed in the charqueadas of Rio Grande do Sul increased by 90 percent; the increase in Uruguay for the corresponding period was of some 300 percent.[72] This commercial crisis had its roots more in overproduction than in any diminution of consumption.

The crisis, however, was real. In 1862 seven mercantile houses in Rio Grande and Pelotas "crashed."[73] The provincial treasury described the state of the trade in ranch products as "apathy close enough to paralysis."[74] The market for salt-beef had grown so restricted that some charqueadas in Pelotas and saladeros at Montevideo were using cattle only for products other than their meat. This last they were burning.[75] Charque was clearly being overproduced in relation to its markets. As commercial competition became more difficult, some provincial deputies had become especially conscious of the need to extend the markets for charque and its related products.[76] The farsighted were already looking toward a different basis for economic growth. As early as 1854, provincial president Sinimbu had argued that ranchers needed to improve their breeding techniques with a view to the future prospects offered by European markets.[77]

Within the broader southern South American ranching region, the century had opened with Rio Grande do Sul in a privileged position. How long that lasted is not easily established; the wars after 1835 brought mixed results for the ranching economy that defy generalization. The results of a return to peace during the 1850's are clearer. Rio Grande do Sul began to confront competition of a new intensity, especially from its southern neighbor, in marketing traditional ranch products. Although financially dependent on Brazil, Uruguay started to gain the trappings of a nation for the first time during the 1850's. It organized to compete, and Rio Grande do Sul was not well placed to meet this rivalry. By 1860, the growth of the herds under peace had led to a temporary saturation of the traditional markets, accompanied by a commercial crisis. This in turn contained the seeds of modernization, with reformers anxious to bring Europe closer to the daily lives of their region.

Association and Organization

$\mathbf{F}$OLLOWING JOHN WALTON'S interpretation of the historical geography of British agriculture during the eighteenth and nineteenth centuries, the diffusion of innovations in South American ranching might also be seen as a prolonged struggle "of attrition between the forces of cautious, eclectic experimentation on the one hand, and those of custom and inertia on the other."[1] A meaningful discussion of that struggle hinges on the specific factors conditioning the chronology of adoption of particular innovations; the diffusion of each innovation is examined separately in this and the following chapters.

The present chapter takes the theme of the emergence of association and organization. It begins by looking for the roots of innovation in the form of agricultural societies and their written literature. It then examines the impact of these societies in fostering the common interests of modernizing ranchers, some of which were expressed through exhibitions and legal codes.

As something that ranchers eventually saw the need to lobby for, rural credit has a close link with rural codes. However, in Rio Grande do Sul innovations on the estâncias of the Campanha and the subsequent refinement of ranch products for the market were both unsupported by any ready access to cheap credit. No development bank in the modern sense emerged during the period under study. The general absence of rural credit, an important part of the explanation for the slow transformation of the Campanha, is examined in Chapter 7.

During the period 1850–1920, politics provided some clear punctuation to the modernization of ranching, especially through the interruption of routines. Any effort to understand the patterns of ranch modernization in the Campanha has to address the interdependency of the region with Uruguay. This was important even before 1850. For example, although Portuguese and Spanish landowners had often been enemies, Tulio Halperín-Donghi has written of the "fruitful contacts" that existed between the Banda Oriental (Uruguay) and Rio Grande do Sul. Gauchos and Indians had kept an unofficial trade route between Brazil and the Plata open.[2]

In the 1850's, Riograndense landowners were still pushing south, buying up huge tracts of land in northern Uruguay, thereby strengthening their cultural hold over that region. These landowners could not resist becoming involved in Uruguayan domestic politics. Largely at their behest, Brazil went so far as to invade Uruguay in 1864; this, in turn, became the pretext for a Paraguayan invasion of Brazil. During the years 1865–70, when Brazil was fighting Paraguay, the conduct of the war, rather than the condition of their estâncias, preoccupied most Riograndense ranchers. A few of them saw their properties directly affected by the fighting; extensive robbery of livestock accompanied the Paraguayan invasion of the western Campanha in 1865. However, this was over a limited area. The Paraguayan War was much more important indirectly, especially in leading to the questioning of tradition throughout Brazil. As part of the restlessness that followed the war, some of the weaknesses of "traditional" ranching in Rio Grande do Sul were drawn to broader attention.

The conclusion of the Paraguayan War provides a definite landmark for ranch modernization throughout southern South America. During the early 1870's, specialist societies catering to forward-looking ranchers were founded in Argentina and Uruguay. Innovations began to take hold on the ranches—in breeding and fencing, for example. These innovations invariably came earliest in Buenos Aires Province and southern Uruguay. After approximately 1870, a modernizing frontier began to diffuse northwards through Uruguayan territory toward the Campanha. Riograndenses who owned land in northern Uruguay stood in the vanguard of change for the Campanha. The important contraband route of the colonial and early national periods had become the leading diffusion channel for Campanha ranch modernization. There was, however, one key difference: this time the impetus for change came from the south and not from Portuguese ambition pushing toward the Plata, in the quest for more temperate land.

As the nineteenth century neared its close, the national political characters of Brazil and Uruguay appeared to be moving in different directions. While the overall trend in the latter was centralization, Brazil's Old Republic brought a more decentralized system. This was only established in southern Brazil with great conflict, however. In the last decades of the Empire, sharp party tension already marked Rio Grande do

Sul; after the disruptions of the Paraguayan War, in Fernando Uri-coechea's phrasing, "the patrimonial charm had been broken."[3] The Campanha became a stronghold of Liberal Party politics, taking a critical stand against policies of centralization and winning useful concessions in such areas as tariff policy. As a recent scholar of Riograndense politics has aptly pointed out, Campanha elites placed their faith in an ascending organization within a falling regime.[4] With the Empire's collapse, differences of ideology and personality between federalists and positivist-centralists (those favoring a parliamentary system of state government over a presidential one) degenerated into a major civil war between 1893 and 1895. Research on the structural causes of the war remains limited, but Silvio Rogério Duncan Baretta's interpretation based on the differential regional effects of market expansion, one of the stresses of modernization, provides useful insights. The war had a clear regional basis within Rio Grande do Sul, with the federalist cause finding its core support within the Campanha. Elites there sought to extend and protect what they had gained under the politics of the late Empire. Like those of the Far-roupilha, the causes of this event, which was very important for the political history of Brazil, still need further research. However, it is clear that the climate of total insecurity in the state during this long struggle was not at all conducive to ranch intensification. Turmoil again interrupted the diffusion channel from Uruguay to the Campanha. War dealt a rude blow to Riograndense ranchers' early efforts at improvement.[5] Much of this war was fought on the Campanha grasslands. The guerrilla character of the fighting brought widespread destruction to those ranches caught up in its theaters, as soldiers tore down fences and removed livestock. Political instability also marked Uruguay in this period, culminating in the unsuccessful Saravia Revolt of 1903–4, a late, desperate challenge to central authority.

Like the Paraguayan War, the civil war of the 1890's provided an impetus for renewal in its aftermath. Rio Grande do Sul emerged from its conflict with state politics under the rigid control of the Partido Republicano Riograndense (PRR), which dealt with broader constituencies than Campanha ranchers, notably urban ones.[6] Even though regional politics were not directly favorable to Campanha interests, the modernization of ranching took hold more firmly under the authoritarian government of the PRR, especially after 1905, when stability returned to Uruguay. The historical geography of innovation in the ranching economy broadly reflected the chronology of these major political struggles.

landowners became closely linked with animal performance. South America shows parallels to the European breeding rhetoric analyzed in recent historical writing.[4]

In Rio Grande do Sul, selective breeding during the period 1850–1920 fell into three phases. Before the Paraguayan War (1864–70), breed improvement was associated with a select band of innovators, most of them rich or eccentric. Without the external stimulus of immigrants interested in raising sheep, the Campanha largely missed the important wool boom that laid a basis for ranch modernization around the Río de la Plata. A second phase lasted from around 1870 through to the civil war of the 1890's. Toward the end of this period, the economic incentives for improving the breeds of cattle began to intensify. Improved livestock only began to be generally adopted in Rio Grande do Sul, as in the peripheral parts of the Río de la Plata, towards the end of the nineteenth century. Much of what took place after 1895 in the Campanha can be viewed as a ripple effect of the changes taking place further south, the slow advance of the modernizing frontier. However, the breeding patterns in Rio Grande do Sul were not simply a later replication of those further south. The transitional character of the physical environment in the Campanha called for a greater degree of experimentation (and therefore risk) with acclimatizing European animals than did more temperate Buenos Aires or southern Uruguay. In the breeds selected as a base for crossbreeding cattle, some original thought emerged from the Campanha. For example, the Devon played a stronger role in Rio Grande do Sul, at least in part as a result of the efforts of Assis Brasil to diffuse the breed.

BREEDING BEFORE 1870

Prior to the 1850's, references to even the idea of stock improvement are sparse in Rio Grande do Sul and usually wholly negative in tone.[5] By the 1830's, there is limited evidence that the most enlightened ranchers were beginning to take some interest in breeding. Even within the "traditional" ranching system, there was room for selection of animals showing specific characteristics.[6] The very unstable political condition of the province from 1835 to 1852 checked any such enthusiasm. But the return to peace brought strong grounds for a renewed interest in stock management.

Nonetheless, the general pattern of activity was not promising. The actual estância practices can be gauged from correspondence received by the provincial presidents from some of the town councils during the

1850's and 1860's.[7] Moreover, a random sample of over a hundred rancher inventories examined in detail, drawn mainly from the counties of Alegrete, Bagé, and São Gabriel, turned up very few explicit references to the presence of improved breeds of livestock during the entire second half of the nineteenth century. While there are several cases where the valuations specifically identify horses as fed on maize rather than pasture alone, there is particular mention of improved animals in only three cases during the imperial period.[8]

The reports of the provincial presidents during the same years, however, reveal that the picture regarding breeding was not as bleak as some of the county reports had been inclined to paint it. For example, in 1852 Vice-President Luís Alves de Oliveira Belo indicated that private efforts at improvement had introduced English thoroughbred horses, brought in from Rio de Janeiro and the Cape of Good Hope, to the province.[9] Sinimbu noted in 1854 that some of the ranchers from Cruz Alta in the Serra were already importing improved bulls from São Paulo and Minas Gerais, as well as the improved strains of creole horses known as pampas, probably from São Paulo. He also claimed that the best work in this regard was being done by Commander Manoel Ferreira Porto at his estância Curral Alto in Triunfo, which already contained various breeds of cattle, horses, and sheep.[10] But none of the specific efforts documented by Sinimbu were taking place in the Campanha.

Sinimbu's same report of 1854 described the counties of São Borja and Bagé as places undertaking merino sheepbreeding. São Borja, in impoverished Missões, was an unlikely locus for innovation. The presence of several flocks of improved sheep there can probably be attributed, at least in part, to the enthusiastic work of the French naturalist Aimé Bonpland, who spent years living around São Borja. Bagé undoubtedly benefited from its location as one of the Campanha counties closest to sheep breeding in northern Uruguay.[11]

Before the Paraguayan War, improved breeding was exceptional enough in Rio Grande do Sul to attract interest from most educated people. Their interest has left a documentary record which allows for the detailed reconstruction of the principal experiments. Many of the ideas outlined by successive administrations about how to improve livestock in the province in the second half of the century stem from General Andréa's energetic occupancy of the presidency of Rio Grande do Sul during much of 1848–50.[12]

During his administration, Andréa gave official expression to his interest, taking steps to remedy the state of the degenerated breeds. A provincial law passed during 1848 established the conditions for setting up a breeding station (*coudelaria*).[13] The rationale for the law was to spend public money in order to stimulate private initiative. In 1848, the government did not see the need for direct involvement in improving the ranching economy.

Andréa's report to the legislative assembly for 1849 included his views as to how the improvement of the regional livestock might take place. He noted that he had looked in vain for materials on stock breeding in the national capital and thus had sought their acquisition in the countries known to pay attention to such matters. His immediate aim was to collate this information in a condensed breeding instruction manual for distribution to the ranchers of the province. After the ranchers had received these pamphlets, Andréa visualized ranching as a process of gradual improvement of the livestock. Under the stimulus of prizes, the local ranchers would bring in pairs of animals from foreign countries (presumably he had Europe in mind). He thought eight to ten years of consistent effort and expense would likely provide the answer to the problems of stock improvement.[14]

Andréa's thinking on breeding proved idealistic in the regional milieu. In spite of this, his successors in the provincial presidency followed through with some of his interest.[15] Most of the official efforts were expended on a significant effort to develop sheep breeding in Rio Grande do Sul, in an attempt to replicate the boom in wool exports already beginning in Buenos Aires Province and southern Uruguay. Official interest in improving sheep came not only from a recognition of their economic utility. The management of sheep was more labor-intensive than that of cattle; there were probably also elements of an early job-creation program in the government concern of the 1850's, when the province sought to occupy some of the people displaced from their homes during the wars of the previous decades.

In the Plata, sheep breeding was associated primarily with foreign-born minorities. Much of the initial impulse came from British merchants who had invested some of their capital in land. Serving as a successful mechanism for the accumulation of capital, by the middle decades of the century sheep appealed strongly to a diverse band of immigrants. Basques fleeing wars in Spain and about fifteen thousand Irish leaving the destitution of Ireland after the potato famine of 1846 are the groups chiefly associated with the history of sheep in Buenos Aires.[16] Uruguay emerged from the destruction of the Guerra Grande with a serious shortage of sheep, but it shared in the wool boom.[17] Sheep were the focus of much attention during the ensuing decades. Specialized works extolling the economic benefits of sheep breeding in Uruguay were published in England.[18]

There was no equivalent published literature coming from Rio Grande do Sul. Nor were foreign minorities with means present in any number to act as catalysts for change on the Campanha grasslands. This said, there were some genuine sheep enthusiasts in Rio Grande do Sul, although their contribution to the wider adoption of improved breeding was to prove marginal.

A New Industry for the Province. In his report for 1852, Vice-President Oliveira Belo noted that he had looked into the possibility of bringing sheep from Montevideo but found that the animals there were not in the strict sense pedigree. He turned instead to Europe through the agency of Friedrich Schmidt in Hamburg (the Brazilian emigration agent in that port), charging him to send from Saxony 50 pure merino sheep along with 10 rams and a shepherd "versed in the treatment of such animals."[19] Since this experiment in breeding was done at provincial expense, very detailed information survives about the project. The province was so unprepared for this enterprise that it initially imported from Europe not only the technical expertise but even the hay.[20]

The presidential report to the legislative assembly for 1853 included a whole section about the sheep-breeding experiment under the subheading "New Industry of the Province." In the following years, the cost of maintaining the flock was an issue that posed serious administrative problems. The province wanted to hold the original flock intact as "the first element for the foundation of a model estância." Part of the philosophy behind this goal was that it could have an educational value for the "ignorant" capatazes.[21]

Over the next two decades, various provincial governments grappled with the questions of how to administer the imported sheep. By 1854, the flock had already almost doubled in size. The aims that the presidency sought to achieve with it had also expanded. Not only would the flock serve as the first stage in the establishment of a model ranch, it would also provide a means to redress a social problem affecting the public purse: the education of orphan children. Teaching children to be shepherds was seen as a highly suitable form of practical education for the poor, a view that left little ground for optimism about their social ascendancy in the region.[22]

At this stage, the flock was still housed in provisional accommodation. By 1855, the president received authorization to contract out the care of the flock to a "rancher of recognized zeal and honesty" and to dismiss the shepherds who had accompanied the sheep from Germany. Instead, Vice-President Oliveira Belo bought a small farm, the Chácara das Bananeiras, in the environs of Porto Alegre. Part of his justification for this expense was concern over the feeding of the flock. More than two years after its introduction to Rio Grande do Sul, the flock of pedigree sheep was still dependent on old hay imported from Europe; in the *chácara* it might be possible to grow planted pastures.[23]

By 1856 the legislators of the province were growing very restive about costs. Part of the stimulus behind growing pasture, for example, was to cut down on the cost of feeding grain to the sheep. During his presidency, the Barão de Muritiba resolved to separate part of the flock. He wanted it moved to the Chácara de Santa Thereza, where the shepherd was to try reducing the grain ration. This president also considered that the flock would contribute little to the pastoral industry of the province unless the expenses of the breeding experiment could be lowered. Rather than continue working to keep the flock pure, the Barão de Muritiba recommended crossbreeding it with the local creole sheep.[24]

Behind the theoretical considerations of the values and disadvantages of sheep breeding lay the problems of daily flock management. The provincial government's idealism was weakening. By 1857, although the nucleus of the flock was to be kept at the Chácara das Bananeiras to guard against degeneration of the breed, a large portion of it (120 animals) had been distributed in lots to various people, many of them among the provincial elite.[25] In 1858, the province rented out the nucleus of the flock at Bananeiras along with the property.[26]

President Joaquim Antão Fernandes Leão's 1860 report set the sheep experiment in a broader context; he claimed that there were 52 purebred sheep in the experimental flock and 17,219 "in the different parts of the province," although these were not strictly pedigree sheep. The number of ordinary sheep in Rio Grande do Sul was placed at 359,293.[27] Even allowing for gross underestimation of numbers, this bears no comparison to the 14 million sheep to be found in Argentina at the same time.[28] By 1864, the provincial government also viewed the management of the provincial flock by lease as unsatisfactory.[29]

Provincial merino breeding failed for numerous reasons>The government was not equipped for a breeding experiment when the sheep arrived from Europe; contingency planning characterized the initiative from the outset. Importing hay from Hamburg to a region with millions of hectares of grasslands was as divorced from good economic sense as the shipping of shirt-collars to London laundries on the part of the Brazilian elite. The breeding project was badly managed, in part from lack of experience but also from lack of continuity, a result of the Empire's frequent shifts of provincial president. Most of these officials were so wary of expense that they dithered in their policies for management. Ultimately, the sheep experiment was a telling example of the continuing gulf between the official world and the character of rural Brazil.[30] When Vice-President Oliveira Belo (a politician with a considerable local knowledge of Rio Grande do Sul) argued in 1855 that the region's most intelligent ranchers were those who spent least time on their lands, he summarized the official view of the matter with remarkable economy.[31] In its concern for breeding, the provincial government failed to carry its message beyond a narrow section of the elite. Those who were involved in early efforts to improve livestock in Rio Grande do Sul almost always had interests around the coastal towns. Apart from members of elite families carrying such names as Ferreira Porto, Gonçalves Chaves, or Pinheiro Machado, the only others who devoted significant energy to early breed improvement were the handful of European-born or "Europeanizing" idealists. The provincial experiments with sheep achieved little. In addition, they soured the idea of state initiative in breeding for the remainder of the Empire.

Early Private Initiatives. There were rare individuals interested in breeding sheep who did not require any special stimulus from the province. For example, Alfred Duclos, a Frenchman married to a cousin of Sinimbu, was raising pedigree sheep around 1850 near Triunfo in the Jacuí Valley.[32] Duclos provided Sinimbu with a detailed account of his own efforts at sheep breeding. He maintained that the flocks closer to the coast were more robust; breeding in parts of the interior required the use of salt, a relatively expensive commodity. Duclos also presented his own formula for making a medicine to cure sheep diseases based on the local plants of the province. Although he castigated the population of the interior for giving too wide a currency to the word "peste," his own approach to breeding appears to have been pseudoscientific.[33]

While Duclos was breeding merino sheep on his estância near Rio Pardo during the 1850's, Aimé Bonpland, a fellow Frenchman, was extremely active in the same regard in the area around São Borja.[34] As an individual of worldwide scientific fame, Bonpland was an interesting potential catalyst for change in Rio Grande do Sul. His interest in merinos ran back to his work at Malmaison in Napoleonic France. In the 1830's, Bonpland pioneered the breed in Corrientes in a brief but large-scale experiment. Political instability there subsequently drew him into Rio Grande do Sul. Through journeyings around southern South America, Bonpland provided a clear channel for the diffusion of ideas about breeding. His register of correspondence reveals that he was a link in a network of merino sheep breeders, in contact with improving ranchers in Argentina and Uruguay, and prepared to help those in Rio Grande do Sul who showed interest.

In the cultural milieu of mid-century Rio Grande do Sul, however, the interested audience was tiny. Bonpland's endless interest in economic diversification spoke of a mentality that was out of step with the surroundings. By the time he was promoting merinos in Rio Grande do Sul, Bonpland was an old man of precarious means. While he had accumulated a long history of economic ventures in South America, it was one marked throughout by bad luck. None of this should imply that he was devoid of energy by 1850; on the contrary, his continuing zest for thoughtful observation is revealed very clearly in his letters to the rancher Antônio Rodrigues Chaves, to whom he was trying to sell sheep.[35] Yet a migratory life, partly chosen and partly imposed (by political adversity and by the need for periodic trips to the major coastal cities in order to collect his French pension), diffused Bonpland's energies and reduced his potential impact.

Merinos could only be acclimatized in southern Brazil with a great deal of careful management, as both Duclos and Bonpland made fully clear. Both these enthusiasts were aware that without constant attention, merinos in Rio Grande do Sul were very susceptible to disease. Even in the Plata, there was a deep-seated prejudice against sheep, difficult animals to control on the open range. In the Río de la Plata, sheep were strongly associated with tenant farming. In the earlier stages, the landowners used a profit-sharing system (*mediería*) that was highly favorable to tenants, in an effort to heighten their sense of responsibility. In the mediería, the shepherd responsible for a flock received a share of the wool clip, although not always as much as the half which the term implies.[36] It took the demonstrated success of immigrants for the gauchos of the Río de la Plata to show any regard for sheep; in the Campanha, newcomers were not present in any number to make this demonstration.

The wool boom of this period provided a very important stimulus for ranch modernization in the Plata. For example, concern for the quality of the fleece provided an important early motive to fence the ranges. Although they often shared the same pastures, sheep called for an entirely different cultural complex than cattle. Cattle might survive the rough handling of the lasso, but sheep would not; they required the specialized care of shepherds, who could not accomplish their routines entirely from the saddle, especially during shearing and lambing. A sharp spatial separation of the sexes marked the cattle cultures of the pampas. The central equestrian tasks of the cattle economy (such as roundups and the breaking of horses) made no room for women. Sheep extended the scope for family labor. For example, women were extensively involved in shearing. In short, where sheep took hold, they offered an important challenge to the traditional gaucho cultures on multiple fronts; in Argentina, the wool boom laid essential groundwork for the country's impressive agrarian development of the early twentieth century. Before 1870, sheep challenged the gaucho way of life in the Río de la Plata in a manner that was absent from the Campanha.

BREEDING DURING THE LATE EMPIRE (1870–89)

The Paraguayan War did much to expose the inferior quality of the breeds in Rio Grande do Sul to a broader audience. The aftermath of the war brought an intensified interest in the idea of selective breeding. However, while the idea of improved breeding took firm root in this period, the practice remained on a modest scale.

A leading critic of Rio Grande do Sul's creole horses was Louis Gaston d'Orléans, the Conde d'Eu, the Emperor's French son-in-law. He attributed the limited stamina of the creole horses to their being fed nothing but grass, noting that it was necessary for those on urgent business to travel with a minimum of three mounts. Compared to the splendid ornamentation of the saddlery of the gaúchos, he also found striking the lack of attention paid to the care and feeding of horses. While an expensive horse cost the equivalent of around 10 pounds sterling, the decorated harness (*arreio*) was usually worth between 60 and 100 pounds. Yet spending money to sustain horses with grain was out of the question.[37]

Despite these criticisms from a powerful source, the quality of the horses does not appear to have improved much immediately after the war. When the Visconde de Pelotas, minister of war in the Saraiva cabinet, announced in 1880 that he was establishing a breeding station in Rio Grande do Sul (within the Invernada Saicã), a local commentator emphasized that the quality of the horses in the province was "noticeably degenerating through the inertia and ignorance of our ranchers."[38] President Henrique Francisco d'Ávila echoed this concern when he said in 1880 that it was impossible to find one good horse among the province's foals and hence asked the Barão de Ijuí to look around the Río de la Plata area for the best possible creole horses to stock the breeding station.[39]

While there were still complaints about the quality of the local horses, interest in breeding did intensify after the Paraguayan War was concluded. Like earlier struggles in the region, the war had interrupted the rhythm of work on the estâncias and provided an opportunity for renewal. For example, in 1871, Coronel Augusto Pereira de Carvalho established his Estância Sant'Anna in Livramento with six purebred Shorthorn bulls and six cows. These were very likely the first "pure" Shorthorns brought into the province.[40] During 1880, references begin to appear to the work of G. W. Armishaw and Company, which imported cattle and horses directly from England.[41] Commenting on the opening of a hippodrome at Porto Alegre in 1880, one observer could see that this represented more than mere entertainment. Interest in racing horses would soon prove a powerful motor for improved breeding.[42]

By the late Empire, provincial politicians were showing more interest in the nature of the animal breeds, motivated mainly by worry over competition. For example, deputy Rodrigo de Azambuja Villa Nova claimed in 1885 that the local creole cattle were visibly degenerating from year to year for want of good-quality bulls. In addition, he argued that Rio Grande do Sul's inferiority to Uruguay in ranching was well recognized by now.[43] In 1887, local leaders debated the merits of a proposal by the young politician Joaquim Francisco de Assis Brasil (first elected to the provincial assembly in 1884) for a breeding station in Rio Grande do Sul.[44]

During the early 1880's, interest in improving the state of the creole cattle had begun to broaden among the ranchers of Rio Grande do Sul.[45] They were strongly influenced in this by the success of the animals crossbred with Shorthorn bulls by Carlos Reyles, a leading Uruguayan breeder. Reyles had been engaged in this venture since 1861, and his mestizo cattle yielded some 25 to 30 percent more in their carcasses than the creole kind. Pelotas-based charqueadores sought out these improved cattle.[46] By 1887, Reyles noted that he could not keep abreast of the demand for high-quality crossbreeds (*mestizos finos*) from both the estancieiros and the charqueadores of Rio Grande do Sul.[47]

Sheep, on the other hand, continued to be a neglected part of Campanha estância life in this period. Even in the late 1880's, the authorities at São Gabriel regretted that sheep did not receive the attention in their county they judged them to deserve.[48] An explanation for this lay in part with the rejection of mutton, considered "almost unfit for food" by the local inhabitants.[49]

Although crossbreeding for the Pelotas charqueadas offered ranchers of northern Uruguay and the Campanha a glimpse of an economic rationale during the 1880's for improving their cattle, the practice remained confined to a small minority in the region. For example, the political chief of the Uruguayan frontier department of Artigas, one heavily marked by a Brazilian landowning presence, noted in 1889 that of the hundreds of thousands of animals present, he could identify only 1,505 cattle, 162 sheep, and 12 horses as of improved breed, whether pure or mestizo.[50] The diffusion of improved cattle through northern Uruguay to the Campanha, already slow, was interrupted during the early 1890's by civil war in Rio Grande do Sul.

BREEDING DURING 1895–1920

After the civil war crossbreeding began to diffuse more broadly in Rio Grande do Sul. It was stimulated in the case of cattle by the higher prices that the charqueadas were able to pay under tariff protection. Sheep finally became important in parts of the state. On some of the ranches where the recent civil war had decimated the herds of cattle, the estancieiros turned to sheep as a cheaper animal for restocking their pastures. Most of the livestock bought for selective breeding came from the Río de la Plata.[51]

Cattle. The record of adoption of the improved breeds of cattle in Rio Grande do Sul is extremely scanty. Although British consuls used language which demonstrates that they had only the vaguest sense of what was taking place on the ranches, Consul E. J. Wigg was correct in surmising that most highbred cattle were introduced from the Río de la Plata.[52] After 1895, the *cabañeros*, the specialist breeders of Argentina and Uruguay, competed to supply the Riograndense market for breeding stock. Environmental similarities between the Campanha and much of Uruguay gave the Uruguayan breeders an advantage over the Argentineans in this trade.

TABLE 5.1

Cattle Imports into Rio Grande do Sul, 1908–12

	From Uruguay	From Argentina	From Minas Gerais	From Europe	Total
1908	129,645	19,277	8,010	1,150	158,082
1909	146,074	21,970	9,650	1,250	178,944
1910	127,137	20,504	11,060	1,350	160,051
1911	145,987	24,992	13,890	1,350	186,219
1912	164,503	34,815	16,125	1,410	216,853
TOTAL	713,346	121,558[a]	58,735	6,510	900,149

SOURCE: *AALRGS* (1914): 32.

[a]Adjusted from 131,588 in the original document.

In the period before World War I, the breeding effort in Rio Grande do Sul was remarkably consistent (Table 5.1). Cattle brought from Europe and Minas Gerais, a part of Brazil strongly associated with the Zebu, were almost certainly introduced for breeding. The much larger numbers from Uruguay and to a lesser extent Argentina probably also conceal some arrivals for slaughter. Such raw statistics alone give no indication of the quality of the animals involved.

Of approximately 300,000 cattle at Itaqui in 1911, at least 13 percent were described as of the Hereford and Shorthorn breeds.[53] At Livramento, almost all of the ranchers engaged in some kind of improved breeding by the First World War. Coronel Augusto Pereira de Carvalho's ranch was distinctive for the variety of animals imported "directly from Europe."[54] For example, there were 2,000 Shorthorn cattle at a high level of crossbreeding, 250 of them purebred.[55]

The establishment of herd-books, genealogical registers of animals, usually trails behind practical experiment. Again, this innovation came late to Rio Grande do Sul. Herd-books were first established in the Río de la Plata countries during the late 1880's.[56] Rio Grande do Sul's first appeared in 1906 and by 1915 had registered only 301 animals.[57] This was on a scale totally different from that of the Río de la Plata. By the end of September 1917, the cumulative registration of cattle in the Argentine herd-book stood at 135,434 head, 77 percent of which were Shorthorns; over ten thousand cattle were being added each year.[58] Uruguay had registered 18,087 cattle in its herd-book by 1914.[59] In highbred livestock, the gap between Rio Grande do Sul and the Plata was clearly enormous. The fact that some Riograndense ranchers chose to

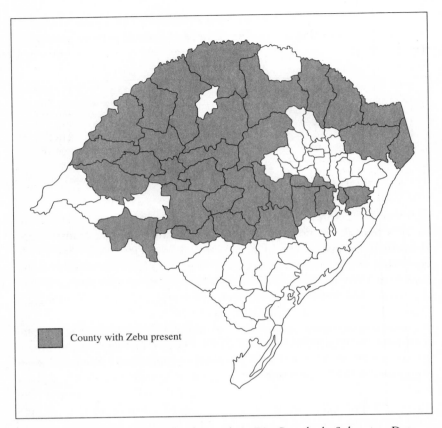

Fig. 5.1. The southern limit of Zebu cattle in Rio Grande do Sul, 1914. Data from State of RGS, *Relatório de Estatística, 1914*, pp. 173–74.

register their pedigree livestock in the Plata rather than in Brazil may have exaggerated it slightly.[60]

The selection of appropriate pedigree bulls for crossbreeding was hardly automatic; it included a fair degree of trial and error. Many ranchers flirted with the idea of reducing this risk by breeding from the Zebu (*Bos indicus*), an animal easily acclimatized on Riograndense grasslands. There was some rationale for this on the poorer, subtropical pastures of the Serra, but the meat of the Zebu did not find favor in international markets. In the Campanha, where it was feasible to improve the creole cattle with European bulls, most ranchers stayed away from the Zebu. In 1914, only one Campanha county along the Uruguayan frontier (Livramento) had adopted the Zebu at all (see Figure 5.1). During the First World War, the breed was in retreat in the Campanha. Those Campanha ranchers who persisted with it brought a very negative response from some of their colleagues.[61]

One of the Campanha rancher's most urgent problems was finding the British breed that worked best, in terms of both the market and local environmental conditions. In Argentina, already shipping meat to Britain, the Shorthorn was the favorite breed, as the most capable of rapidly laying on the fatty beef that British consumers demanded.[62] With careful attention to its forage, the Shorthorn could usually outperform any other breed. However, where the South American rancher still relied predominantly on natural pastures, as was the case in the Campanha, the Shorthorn was a dubious choice.

Average weights achieved by British beef breeds in their domestic settings were not likely to be replicated throughout the different parts of southern South America. Ranchers in the Campanha watched the greater breeding experience reflected in the results of the expositions in Argentina and Uruguay with interest. For example, the average weight of the Shorthorns shown at Paysandú, Uruguay, around 1910 had been 642 kilograms, against the Devons at 658. In England, the results were 1,053 for the Shorthorn against 826 for the Devon. These results helped to justify the use of hardier breeds, such as the Devon or Polled Angus, for crossbreeding. The contemporary literature often describes the Devon and the Polled Angus as "rustic." This meant that they could forage successfully even on the open range and did not succumb to the summer heat, to the first signs of drought, or to the lack of winter forage.[63]

By the beginning of World War I, a polemical literature on stock breeding was starting to develop in Rio Grande do Sul. Delfino Riet, vice-president of the Breeders' Union, observed that Shorthorns and Herefords thrived in the best pastures of Uruguay and in Buenos Aires Province but not, in his view, in Corrientes, the inferior Uruguayan pastures, or the greater part of Rio Grande do Sul. Riet advised the more rustic Devon and Polled Angus breeds for Rio Grande do Sul, preferring the latter (as a better beef animal) to the versatile Devon breed.[64]

Following experiments with the Durham and the Hereford, Assis Brasil, on the other hand, argued with passion that the Devon was the breed best suited to the environmental conditions of Rio Grande do Sul.[65] He maintained that animals bred on the Devonian geological formation (lacking in calcium phosphate) would serve best for crossbreed-

Fig. 5.2. *Farrapo*, a Red Polled bull, in Pelotas, c. 1920. From Brazil, *Recenseamento de 1920*, 3(1), opposite p. 416.

ing with a base of the old Iberian creole cattle.[66] His rationale for this was that much of Rio Grande do Sul had a similar shortage of calcium phosphate.[67]

Campanha ranchers of narrower experience than Riet and Assis Brasil undoubtedly found the divided expert opinions on cattle breeds bewildering. Confronted with a lack of consensus, some of them took years to settle on a breed. Many could not resist trying a little of everything available on their estâncias. Thus, in 1917, of the 24 leading cattle ranchers at Alegrete, less than a third of them (seven) worked with a single breed.[68]

The pace of breeding activity gathered real momentum in the decade punctuated by World War I. During the war, Riograndense breeders bought hundreds of thousands of creole cattle from Uruguay, Paraguay, Corrientes, and Entre Ríos to stock their lands and for crossbreeding.[69] By 1918, the proportion of cattle bred in part from the leading British beef breeds was at least 37 percent of the regional herd (2,000,000 Hereford, 1,000,000 Shorthorn, 150,000 Devon, and 100,000 Polled Angus).[70] Assis Brasil had become the owner of the largest herd of purebred Devons in the world by 1922.[71] Despite all the polemic about the virtues of Devon and Polled Angus bulls for Rio Grande do Sul, most of the ranchers there had opted for the two beef breeds most commonly used in Argentina and Uruguay. As in Uruguay (but not in Argentina), the Hereford predominated in Rio Grande do Sul, reflecting the fact that winter forage availability did not compare with that on the humid pampas.

Sheep. Beginning in the 1890's, sheep finally gained significance for some Riograndense ranchers, mainly for wool. As promoters were quick to point out, ranching enterprises that included sheep usually saw financial results earlier than those starting with cattle alone.[72] In the decades prior to the First World War, ranchers were crossbreeding their creole ewes with Rambouillet, Romney Marsh, Blackface, and Lincoln rams, mainly for wool. Pereira de Carvalho, the Livramento cattle breeder, also owned 2,000 Rambouillet sheep of high quality by 1910, and a particular emphasis on sheep was finally developing in that county.[73] Based on the 1920 census data, Figure 5.4 shows that wool zones had

Fig. 5.3. Pedigree Rambouillet sheep on the Fazenda da Vista Alegre in Livramento, c. 1913. From *A Estância* (Dec. 1913): 296.

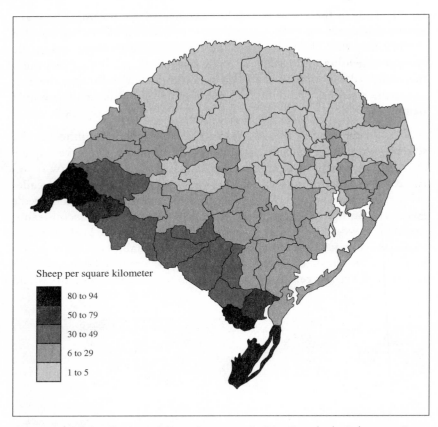

Fig. 5.4. The distribution of sheep by county in Rio Grande do Sul, 1920. Data from Brazil, *Recenseamento de 1920*, 3(1): 415–17.

emerged; by this time, in fact, sheep outnumbered cattle in the southwest and southeast corners of the state.

The southwestern Campanha was the most important wool zone in Rio Grande do Sul; its development was recognized as having followed the extension of the railways into northwestern Uruguay.[74] It is a clear example of the extending reach of the modernizing frontier coming from the south. After 1905, there was a sharp reduction of railway freight rates for wool carried between the Campanha frontier and Montevideo; this intensified the appeal for Riograndense wool producers of using the Uruguayan railways.[75] Wool shipped by rail to the established merchants of Montevideo fetched better prices. In turn, the Brazilian textile industry bought wool lots in Montevideo sent from such Campanha counties as Alegrete, Quaraí, and Uruguaiana.[76]

Contemporary comment on the wool zone in southeastern Rio Grande do Sul is scarce. However, a 1912 description of the leading Jaguarão estâncias shows that Uruguay had played the leading role in supplying breeding stock.[77] Although sheep had diffused widely in some of the counties along the southern frontier after 1890, it was reported in 1917 that ranchers in "certain zones" of the Campanha still considered their pastures unsuitable for them. Delfino Riet, one of the leading Riograndense experts on the subject, attacked this continuing unfounded prejudice in print. Outside some of the Campanha counties confronting Uruguay, the case for sheep was still being pondered in 1920.

Horses. By World War I, the interest in improving breeds had extended to horses. The leading agents for the improvement of horses continued to be sport and the military. Improvements in transport to Brazil's major cities allowed some Campanha ranchers to make a business out of breeding for the turf. These included Assis Brasil, some of whose horses won considerable prize money racing in Rio de Janeiro.[78] The quality of horses also received attention from the federal government on the national stud-farm at Saicã, which ranchers were encouraged to use for improving the quality of their own horses. By 1917 this contained 40 thoroughbred horses. The 22 stallions were mainly of Arab, Anglo-Arab, or English blood.[79]

Breeding moved in stages, from government rhetoric and ill-fated experiment, to various isolated initiatives by individuals, usually foreign-born, to the more widespread adoption of improved breeding practices, often in an attempt to emulate the demonstrated success of Argentinean and Uruguayan competitors. Even so, whether in the case of cattle, sheep, or horses, improved breeding took hold slowly in Rio Grande do Sul, even after the 1880's, when the idea had been firmly implanted. As late as 1920, the number of animals of outstanding quality in the region was probably still very small, except perhaps for racehorses.

Fencing the Open Range

Fencing played a variety of roles in the modernization of ranching, as Domingo Ordoñana, an enthusiast of rural modernization who witnessed the process at first hand in Uruguay, appreciated.[80] Fencing the perimeter of a property gave the owner demonstrated dominion over that land. In a region where few properties had ever been measured, fencing provided a clear symbol of the consolidation of the estate. No matter how close the management of any large estate on the open range, it always showed elements of collective property. On estancias of thousands of hectares, it was impossible to control the behavior of neighbors and strangers all the time. Before the estancias gained perimeter fences, there were problems with dispersion of herds, either on account of seasonal droughts or because herds stampeded when surprised by strangers. Fencing put an end to most of this; in Raúl Jacob's phrasing, wire represented a zealous guardian of private property.[81]

While perimeter fences helped to define property, subdividing an estancia with internal fences gave the owner dominion over the soil and allowed for intensification of production. Fences allowed a ranch to carry more animals per unit area, an important consideration on the estâncias of the Campanha, where a tendency to overstock had been a feature of their "traditional" character. On properties clearly demarcated and protected from the intrusion of animals belonging to neighbors, ranchers could now engage in selective breeding on a much more scientific basis. The internal subdivision of the ranches into fenced paddocks changed estancia management radically. Paddocks helped particularly with sheep, where the concentration of flocks had always been vital to their management. But paddocks also helped with cattle. Even with natural pasture, they allowed for a greater degree of management (ranchers could hold back grasses and adjust grazing densities), enabling a given area to carry greater numbers. In Uruguay at least, where the internal division of ranches was sometimes termed "the Platine system of ranching," the distinctive features included not only the presence of internal fences but also the cultivation of forage crops.

Technology is rarely neutral in its economic and social implications. The changes in ranching technique that accompanied fencing had major labor implications. Fencing affected relations between landowners, but especially between landowners and their staff. For example, perimeter fences cut down on the need for labor; boundary peons (posteiros) and searchers for strays (recrutadores) were no longer required. Instead of holding the rodeios necessary to round up animals on the open range, ranchers built wooden corridors (bretes) within their small corrals. By confining animals in the bretes, the staff could work on them at a fixed site; this economized on labor costs, since it required fewer peons than the extensive rodeios of the "traditional" ranch. Having fewer peons greatly reduced the need for workhorses, since on the open ranch each

peon had kept his own string of five or six creole horses. Grass once devoted to feeding these workhorses could now raise extra cattle or sheep
for sale. This streamlining of ranch management, a common feature of
the "improved" system of ranching, was an important index of modernization in its own right.

The best "traditional" estancias had natural boundaries for the control of livestock. Where they did not, ranchers tried to use other forms
of boundary, such as planted vegetation or stone walls. In Rio Grande
do Sul, where stone was more readily available than on the humid pampas, there were extensive walls on some of the estâncias.[82] However,
while hedges and walls were widely used around the corrals, the expense of construction generally prohibited their broader employment.

The key to the fencing problem lay with industrially manufactured
wire. Ranchers in Argentina had begun to use smooth iron wire on a
small scale as early as 1855. This was very expensive and not completely satisfactory. Iron wire was brittle. The tendency of cattle to rub
against the smooth wire loosened the posts and led with time to broken
fences. Barbed wire was first patented in 1860, but there was much experimentation before it became useful. The technical obstacles to the use
of wire fencing had been removed by about 1875 with the refinement of
wire made from tough, high-tensile Bessemer steel, to which barbs were
added by machine.[83] The circumstances and implications of adopting
this technology varied considerably between the core of the southern
South American ranching region and some of the peripheries. While the
pampas of Argentina are the best-studied case, they are generally not a
reliable model for what took place in the remainder of the region. Wire
fitted into the diverse grassland systems of the region in some distinctive
ways.

The earliest serious adoption of wire fencing in South America came
around the Río de la Plata. The process began with the Europeanizing
elite of progressive ranchers (particularly the minority groups so concerned with sheep) in Buenos Aires Province and the southern departments of Uruguay. General adoption of wire was much slower. Fencing
the millions of hectares of South American grassland was a long affair;
for example, while the fencing of Argentina with wire began after 1875,
the task continued until around 1910. According to the estimates of the
Argentinean author Noel Sbarra, between 1876 and 1907 Argentina
imported enough wire to fence the perimeter of the republic with seven
strands some 140 times.[84]

Although it radically changed the management of ranches and provided one of the most dramatic changes to the rural landscape, the process of fencing the grasslands of Rio Grande do Sul has not been widely chronicled.[85] Without an agricultural press to monitor its adoption, sources for the study of fencing are fewer in Rio Grande do Sul than in the Río de la Plata. Wire fencing constitutes another example of the modernizing frontier arriving in the Campanha through Uruguay. The lag behind the Plata in the generalized adoption of fencing was not as marked as with many other innovations, a testimony to its practical values. Nevertheless, the adoption of fencing was slower in the Campanha than in much of Uruguay.[86] Although the territorial character of Uruguay and the Campanha is similar, the circumstances surrounding fencing contained an impressive series of differences, key among them being Uruguay's use of developmental policies, instruments that were absent in the Campanha. In this sense, fencing in Uruguay offers a key to understanding the patterns of adoption in the Campanha.

Uruguay is distinctive within the fencing history of the southern South American ranching region as a whole. It was fenced with wire much more quickly than Argentina, despite that country's superior economic power.[87] Barrán and Nahum have calculated that 64 percent of the parcels of land (*suertes*) were fenced by 1882.[88] In addition, they estimate that 87 percent of this portion of the national territory was fenced in the years from 1877 to 1882 alone.[89] Scholars have viewed Uruguay's homogeneous territory as part of the explanation for this rapid adoption. However, the physical character of the Campanha could also be described as homogeneous, and fencing generally came later there. A more convincing explanation lies with developmental policies.

In the key decade after 1872, Barrán and Nahum estimate that Uruguayan landowners invested the equivalent of two years' production from all breeding ranches in the country in wire fencing, a massive investment of capital.[90] How they achieved this is not completely clear. A mortgage bank founded in 1872 may have provided loans to some of the ranchers.[91] Uruguayan progressives certainly used their Rural Association to encourage fencing. The leading members successfully lobbied the government for tax relief. A customs law of 22 October 1875 granted Uruguayans an exemption from import duty on barbed wire. Although this measure was short-lived, with a financial crisis in 1879 causing the government to reestablish its tax, the duty-free period appears to have provided a powerful stimulus for the adoption of fencing. Uruguay's consumption of barbed wire tripled between 1875 and 1877–78 to more than six thousand tons a year. In addition, the Uruguayan government instituted other developmental policies. Land fenced or land growing forage crops carried a lower land tax (*contribución directa*) after 1876. Reforms to the rural code provided further stimulus for fencing. For example, neighboring landowners shared the cost of putting up common fences after 1879.[92]

Contemporary estimates are scarce, but the few available indicate that the cost of fencing in Uruguay fell quite dramatically during the course of the 1870's. The total cost of fencing the perimeter of a square league (2,656 hectares in Uruguay) fell from 6,000 pesos in 1874 to 3,870 pesos in 1882; in addition, the sharing of fencing costs between neighbors after 1879 further halved this to 1,935 pesos per landowner.[93] Barrán and Nahum have reckoned that a perimeter fence embodied the equivalent of a third of the value of the land in 1874 but represented far less by the early 1880's. For example, from the 1886 inventory of the famous rancher Carlos Reyles, they conclude that the fences at his Estancia Bella Vista in Tacuarembó represented only some 7 percent of its value.[94]

Despite the general decline in costs, northern Uruguay adopted barbed-wire fencing slowly. Part of the explanation for this still lies with cost; according to the calculations presented by Carlos María de Pena, a contemporary author, in 1882 it cost almost three-quarters as much again to fence a *cuadra* (approximately 85 meters square, or slightly under three-quarters of a hectare) with posts and wire in the northern departments of Tacuarembó and Cerro Largo as in the rural environs of the capital.[95] Fencers in northern Uruguay still faced the obstacle of the high cost of moving posts and wire overland from Montevideo. Nevertheless, the developmental measures identified above had made more general adoption economically feasible.

The demonstration of the advantages of fencing in Uruguay was not lost on the ranchers of the Campanha; however, fencing was not solely a matter of economics. Brazilians in northern Uruguay appear to offer a clear demonstration of Friedmann's development proposition that peripheral elites will attempt to resist those innovations they regard as undermining their own authority positions.[96] In this case, they recognized the potential of wire to change the power they held in this region. The fact that the process of fencing was guided from Montevideo had major implications. In the 1870's, Riograndense landowners still controlled most of northern Uruguay. They tended to view fencing and the rural code guiding it on Uruguayan soil as technical symbols of the unwelcome extension of Montevideo's power. On the political level, fencing represented "absolute centralization" for these Brazilians resident in Uruguay, and it challenged their perceived control over the frontier region. That challenge engendered political instability, which in itself acted as a general impediment to fencing in northern Uruguay.[97] A second major obstacle was cultural: fencing brought a threat to their traditional way of life on the open range.

Nonetheless, some of the extraordinary momentum during the later 1870's for land enclosure in Uruguay carried repercussions for the Campanha. Much of the fencing wire employed in the Campanha entered the region as contraband from Uruguay. By its nature, contraband does not lead easily to estimation of quantity, and there is no indication of just how much fencing wire was brought across the Uruguayan frontier into Rio Grande do Sul. But it was important enough for Silveira Martins to use it (in an 1880 Brazilian senate debate) as a key example of manufactured goods entering Rio Grande do Sul illegally from Uruguay, and as one reason that a special tariff should be granted to Rio Grande do Sul by the imperial government.[98] Unfortunately, Silveira Martins, widely known as The Tribune, made his argument using the authority of a commanding presence rather than with statistics.

Nor is a firm record of the quantities of wire imported directly into Rio Grande do Sul easily established. Since Britain was the leading supplier of fencing wire to southern South America, the quantities exported to Rio Grande do Sul might be expected to appear in the annual trade reports of the British consuls. However, at Rio Grande the consuls compiled their trade reports without the benefit of the customhouse import statistics. Hence, as late as 1896, Consul Ralph Bernal could write only vaguely: "It would appear that the wire is imported from England, Germany and Belgium, but the quantity sent from each country is unknown."[99]

The limited statistics available do confirm that fencing wire was becoming important in Rio Grande do Sul during the 1880's. Descriptions of the manifests of ships entering Pelotas between January and July 1880 include those of both British and Norwegian vessels carrying fencing wire from Liverpool.[100] Commenting on the arrival of the *Leibnitz* from Liverpool in 1881, the Montevideo newspaper *A Pátria* noted that the English barbed-wire manufacturers were working almost exclusively to supply the markets of the Río de la Plata and Rio Grande do Sul.[101]

The year 1883 marks the first specific mention by a British consul of wire entering Rio Grande do Sul; it was part of the salvaged cargo of a wreck.[102] In 1885, the newly opened railway from Rio Grande to Bagé carried 1,314 tons of fencing wire into the Campanha, 68 percent of it from Pelotas.[103] There are no further details until after the civil war of 1893–95, when wire must have been in strong demand for effecting repairs to fences destroyed in the fighting. Consul Bernal recorded imports of approximately 838 tons of fencing wire into Rio Grande during 1896 and 1,403 tons in 1897; however, Bernal was unable to provide fencing wire import statistics for Pelotas, where, he admitted, the quantity imported was "far greater."[104]

Although the British consuls at Rio Grande said little about fencing wire imports (and their colleagues at Montevideo said nothing specific about contraband wire destined for Brazil), the few available statistics point to lower quantities of wire in Rio Grande do Sul than in Uruguay. None of the quantities mentioned in British consular reports from Rio Grande compares with the 6,000 tons of fencing wire entering Uruguay in the later 1870's, although it cannot be assumed that all of that wire was used in Uruguay itself. The lack of a reliable statistical record on Rio Grande do Sul's overall consumption of fencing wire works against a quantitative analysis of its diffusion. However, other sources provide insights into the process of adoption in the Campanha.

Some Riograndense ranches started fencing early. According to reports by the county authorities of Jaguarão in 1912, the Fazenda do Juncal received its first strands of barbed wire in 1870.[105] If that date is correct, this was very early adoption, since barbed wire was still in its experimental phase. Surviving accounts confirm that another Jaguarão ranch, part of the Estância Rincão de São João, was also addressing its fencing problem by 1872.[106] Location favored these early Jaguarão fencers. Supplies for their ranches could be brought from Rio Grande or Pelotas by water through the Lagoa Mirim more cheaply than in many parts of the Campanha proper. And those who wished to avoid the heavy import taxes levied at the customhouse in Rio Grande could smuggle expensive fencing wire in from Uruguay, whose territory was never more than a few enticing kilometers away.

In 1879, the imperial government organized the fencing of the national stud-farm at Saicã. This São Gabriel property, used for breeding horses for the Brazilian cavalry, covered an area of approximately 50,000 hectares.[107]

Revisions to the municipal code at Bagé during the 1880's imply that fencing had become a significant matter in that county. In a list of articles sent to the provincial president in 1884 for approval, one provision made for the policing of the estâncias fined owners who herded their animals in neighboring lands without the permission of the landowners; most of the land must still have been open at that stage.[108] Three years later, the council added a series of new articles to the municipal code dealing explicitly with fencing and roads. For example, it made provision for those involved with moving livestock. Where the ranchers fenced both sides of municipal and local roads, they were expected to leave a corridor of fifty meters in breadth for the movement of animals. Gates (*porteiras*) to allow animals out of the corridors of passage in order to rest were also to be provided.[109]

Ranch accounts, such as that of a well-financed *invernada* (fattening operation) that formed part of the Estância Rincão de São João in Jaguarão, testify to the gradual adoption of fencing. São João specialized in buying droves of cattle from the interior of Rio Grande do Sul and from Uruguay, and fattening them for slaughter in Pelotas. In late 1872, the four partners in the enterprise—José Antonio Moreira, João Francisco Gonçalves, and Gabriel and Francisco José Gonçalves da Silva—decided to allocate their profits of the past two years (7:621$564, or around 800 pounds sterling) to fencing their property.[110]

The fencing was entrusted to a German, "Francisco Cheffar."[111] An entry in the accounts on 28 October 1872 records a payment to him to build an "asçude" (reservoir), a new corral (*mangueira*), and fencing. The work and cost of doing the fencing were spread out until at least the end of 1875. It is not clear what kind of fencing was being used, and the accounts are not broken down into sufficient detail that the materials used could provide clues. A contract that "Cheffar" held after October 1875 hints that this was fencing composed at least in part of vegetation.[112] On 30 September 1878, the accounts include record of a contract made with "Cheffar" on 19 July 1878 to look after the fencing "for the third and last year."

The accounts refer to the shipment of fenceposts (*moirões*) in 1880, to the construction of two large reservoirs in 1881–82, to the renovation of 1,512 *braças* (3,326 meters) of "vallo" (fencing probably made from a combination of earth-banks and vegetation), and to the erection of eight *quadras* (1,056 meters) of wire fencing (the accounts use the Spanish word "alambrado" to describe this). There is a gap in the accounts from late 1884 until 1895. In 1895, a considerable amount was spent on labor making earth-banks (vallo), and during the following years there were regular payments for posts, rolls of wire ("roulos de arame"), and labor engaged in the maintenance of the fences and of the reservoirs.

Even on the well-financed Invernada São João, fencing was a process of gradual investment. This was likely the general pattern in Rio Grande do Sul; that impression is borne out by the descriptions of other Jaguarão ranches in a book, published by the council (intendência) of Jaguarão during 1912, which reviewed some of the recent developments in the area.[113] Often, the estancieiros fenced the perimeters of estâncias first, particularly where their properties bounded lands that did not belong to their own extended families. In the latter decades of the nineteenth century, fencing with wire was only beginning. It was a gradual process, one that had not been entirely completed even by 1920.

OPPOSITION TO FENCING

Closing the open range was not without problems. Fencing often meant closing off roads and trails, which was seen to damage the public interest. As the pace of fencing intensified in the early 1880's, animals still moved on the hoof to the coast for slaughter; there were as yet no railways serving the Campanha or northern Uruguay. Poor fencing practice fell particularly hard on drovers, who found their work greatly complicated by the growing maze of wire that was appearing across the horizons.[114] Without railways and with distant points of slaughter, the Campanha of the 1880's had many staunch defenders of the open range. Given the density of animal traffic converging on the major slaughtering center of Pelotas, and the difficult topography traversed by cattle trails through the Serra do Sudeste (see Figure 3.1), it is not surprising that Duncan Baretta's study of the geography of violence in late-nineteenth-century Rio Grande do Sul found violent disputes over enclosures concentrated in the southeastern corner of the state.[115]

As they fenced, ranchers were much more concerned about their private interests than about the collective public good. In 1877, deputy Manoel Lourenço do Nascimento had made a prescient call for the provincial organization of fencing. Basing his case on the obstacles posed by fencing to the successful movement of fattened cattle toward Pelotas for slaughter, Nascimento argued that ranchers were working against the public good. They fenced land used customarily for roads as well as the water holes used by the tropas on their way to the coast for slaughter. In the rare cases where landowners did pay heed to existing roadways, the land they set aside was usually the least appropriate, the most affected by the seasons.[116]

The problems caused by indiscriminate fencing were reflected in the drove-fees paid when cattle reached Pelotas. These fees, long stable at

3$000 per head, rose to 5$000 as the ranchers began their fencing. Even so, deputy Nascimento claimed the work had become so difficult that it was hard to find people prepared to take on the task. Once the animals were delivered in Pelotas, there was no difficulty in finding rented lands to feed and water them, but it was a totally different affair in the interior. Drovers were only able to sustain the animals under their charge by making immense detours.[117]

Nascimento's proposal met with considerable opposition. Some deputies spoke against the need for a provincial law on fencing, arguing that this was a matter for town councils. Nascimento, however, made the case that the councils would have little power to effect change without the support of a law passed by the provincial assembly. He was correct. Lack of concern for public roadways was still identified as a leading problem at Livramento as late as 1909.[118] In Livramento, as elsewhere in the Campanha, the solution to cattle transport problems caused by indiscriminate fencing came only in the years just before World War I, when railways began to carry animals to their points of slaughter on a significant scale.

THE SOCIAL CONSEQUENCES OF FENCING

The adoption of fencing had important social ramifications, since fenced ranches required less labor than the open range. As the work of the estâncias was accomplished within fenced paddocks, this reduced the need for supplementary help in the rodeios. Nor was the role of the posteiros, the peons who kept watch over the estância margins, any longer so vital. In short, wire fencing removed much of the basis for a long-established agregado class.

People displaced from the Uruguayan ranches congregated in misery in what contemporaries unpleasantly referred to as "rat towns" (*pueblos de ratas*).[119] Some 40,000 of a probable rural population of 400,000 in the Uruguay of 1880 were unemployed.[120] In addition, as the rage for fencing began to touch northern Uruguay and the Campanha by the 1880's, it was accompanied by labor displaced from the departments further south, where fencing had been adopted earlier.[121] As the fencing frontier advanced through northern Uruguay, it intensified problems of social control in the Campanha.

Fencing must have also displaced some of the rural population within Rio Grande do Sul. When the imperial government fenced the national stud-farm at Saicã, the Barão de Batoví had informed the provincial president that "a lot of poor people" lived outside of the fenced zone.[122]

In addition to marginalized free peons, there were still slaves on some of the estâncias in the 1880's, but the numbers began to decline sharply after 1883 (see Chapter 6). There is a close correspondence between the beginnings of adoption of wire fencing on a significant scale and the abolition of slavery in the Campanha counties; with the falling labor needs accompanying enclosure, ranchers no doubt saw they could manage without their slaves.

Labor bore the brunt of the social costs of modernization, but there were also ranchers ill-placed to meet the costs of fixed investment represented by wire fencing. For some, poverty was the chief legacy of the civil war of 1893–95, "the bloodiest . . . in Brazil's history," which saw Riograndense ranches become theaters of war.[123] The war was certainly a factor in the current of out-migration to the Mato Grosso of 1900–1906 that Mário de Lima Beck has chronicled.[124] Another stimulus for out-migration was economic. The increased ranch productivity that followed fencing was reflected in steeply increasing land values, even before the establishment of the frigoríficos. In the Mato Grosso, Riograndense landowners could buy a league of land for the same price as a quadra in their home state.[125] A report made in 1909 by the government of Rio Grande do Sul spoke of the Riograndense out-migrants as "conquered by a more advanced civilization" that was gradually invading the state.[126] This "advanced civilization" was composed of Uruguayan ranchers, who were buying up land in Rio Grande do Sul, especially in the ranching zones of the Serra. At half the prevailing average price at home, this land was cheap to Uruguayans, especially for those from the southern departments, where fencing had commenced earliest.[127]

While the migrations beyond Rio Grande do Sul of land-hungry German colonists from the Serra have received extended comment in academic writing, Germans were not the only group to seek cheaper land in other parts of Brazil.[128] The population movements that accompanied the widespread adoption of fencing in the Campanha warrant further research.

There were important regional variations in the degree to which fencing fostered the modernization of the ranches of southern South America. In the Campanha of the 1880's, fencing was adopted without much cultivation and before the emergence of wool zones. Unlike Buenos Aires and southern Uruguay, the Campanha experienced only a weak stimulus to fence coming from the European markets; in the last decades of the nineteenth century in Rio Grande do Sul, the link between fencing and growth in the domestic salt-beef trade was more important. Campanha ranchers were aware of the economic advantages of fencing their properties; however, those advantages were not as strong as in Buenos Aires Province or southern Uruguay. Married to an absence of the brief but important developmental fencing policies enjoyed by their Uruguayan neighbors, this gave Riograndense ranchers short of liquid capital their rationale for making fencing investment gradual.

Planted Pastures

Modern research has viewed pasture improvement as a key issue in southern South American ranch intensification. As Albert Hirschman has observed, improvement in pastures conditions advances in the subdivision of ranches and in fodder conservation. Pasture improvements have probably also been important for disease control, in that more ample and better-balanced feed have reduced the incidence of disease. In addition, as Hirschman has argued, the "more active livestock management consequent upon subdivision and frequent rotation [in itself has eased] disease detection and control."[129]

There was a considerable variety in the condition of the natural pastures of South America even before the ranchers considered improving them. For example, in the early part of the nineteenth century the natural pastures of Rio Grande do Sul supposedly contained around four hundred species of forage plants. Such diversity could look very impressive in spring but equally depressing in a summer drought and after the first frosts of winter, when animals found little to graze.[130] Loss of animals from lack of pasture was never welcome even on the "traditional" ranch, but as the value of livestock increased it began to weigh much more heavily on ranchers. It made no economic sense for ranchers to expose expensive highbred animals to the risks of lack of pasture; secure year-round forage was vital. Where possible, their ideal solution was to plant carefully managed forage grasses. Planted pastures increased the carrying capacity of the range and enabled fattening throughout the year. Of all the types of grass used, deep-rooted alfalfa, termed the "real queen of the pampa" in Argentina, recommended itself the most, whether grazed by cattle or cut to make hay. It was especially good through times of drought, more common in Argentina than in Rio Grande do Sul. In the 1860's, for example, a British traveler noted around Buenos Aires that alfalfa seemed "capable of producing good crops in seasons where the ordinary grasses of the country dry up from want of rain."[131]

While the general potential of alfalfa was clearly understood, the regional challenges of its adoption took longer to emerge. The most progressive ranchers in the Campanha understood the importance of planted pastures well before 1920. As with fencing wire, the circumstances of adoption showed significant differences between the core and the periphery of the broad pampa grassland region.

By the late 1880's, alfalfa was in cultivation on an extensive scale in Argentina.[132] In the monographs accompanying the 1908 census there, most ranches were described as being turned into "alfalfa plantations," whatever their size.[133] The area sown to alfalfa in Argentina peaked in 1918–19 at over 8.7 million hectares, exceeding even the total for the United States at the time.[134]

The broad pattern was very different in the remainder of the southern South American ranching region. In Uruguay, except for the farms growing hay in the southern departments, the area of land devoted to forage crops was still insignificant in the early part of the twentieth century. The *Album Pur-Sang*, a 1916–17 survey made by journalists of the estancias there, recorded that the majority of the establishments did not devote a single hectare to the cultivation of forage crops.[135]

In Rio Grande do Sul, the Paraguayan War had demonstrated the inadequacy of forage. The practice of importing alfalfa from Argentina, started by the Brazilian army, became a trend during the 1870's. Not all contemporary observers understood the rationale for this. The British consul in Rio de Janeiro, for example, argued that the crop should have been grown in Rio Grande do Sul, a judgment easily delivered from the distance of south-central Brazil.[136] By the 1890's Brazil was importing significant quantities of alfalfa from Argentina; the 325 tons of 1870 had risen to 52,476 by 1894 and continued to rise thereafter.[137] Brazilian ranchers solved their forage problem by buying hay from Argentina, cut at the end of the spring and prepared with little concern for its quality.

The literature on rural improvement in Rio Grande do Sul that was beginning to appear toward the end of the nineteenth century displayed an immediate interest in alfalfa. In *Cultura dos campos*, his manual on cultivation, which was first published in 1898, Assis Brasil stressed the point that careful investment in cultivation could bring high returns from the land and railed against the timidity of Brazilian farmers. He considered that where the ranchers of Rio Grande do Sul grew alfalfa with care, it yielded well on the calcareous soils of the region, providing a better forage than the Argentinean equivalent, which he described as "generally prepared with little scruple."[138] Advice was even available at the scale of the individual county; in 1909, ranchers in Livramento were given the details of forage grasses in their local journal.[139]

At the end of the nineteenth century, alfalfa was on record as grown in almost all of the counties of Rio Grande do Sul.[140] It was by no means confined to the Campanha; for example, French colonists around Pelotas introduced the crop to the Caí and Taquari Valleys (areas of German colonization in the Serra).[141] Like the Welsh in Patagonia, some Riograndense colonial farmers found a successful niche market in selling alfalfa to ranchers for forage.[142] Despite the breadth of cultivation, alfalfa was still grown on a minor scale. Around 1915, a mere 20,620 hectares were seeded with alfalfa in Rio Grande do Sul.[143] When Brazil conducted its first agricultural census in 1920, alfalfa cultivation in Rio Grande do Sul was of sufficient interest to illustrate with a photograph, but no data were collected on its extent.[144] Despite the varied manifestations of interest and the extensive use of imported alfalfa, relatively little was achieved with planted pastures in the Campanha even by 1920.

The explanation for the limited diffusion of planted pastures in the Campanha includes environmental, economic, and cultural impediments. Alfalfa's roots can extend meters below the surface of the ground, so that subsoil permeability becomes an important control on the growth of the crop. Throughout the southern South American ranching region, agronomists advised ranchers to break up the ground by cultivating cereal crops for several years prior to planting alfalfa.[145] In the alfalfa zone of Buenos Aires, the soils are light and frequently very sandy, but with a permeable subsoil from which alfalfa can draw plentiful supplies of water and lime. By contrast, in most of Uruguay and the Campanha of Rio Grande do Sul, the soils are compact and difficult to work during the prolonged wet winters, as well as at times of summer drought. Such difficulties with working the ground translated into greater expenses than in Buenos Aires.[146] Although Assis Brasil had noted in the 1890's that there were limited alfalfa crops in Rio Grande do Sul which yielded very well, many soils in the Campanha were of insufficient depth. The mistaken idea that alfalfa depleted the soils of nutrients remained widespread in Rio Grande do Sul.[147]

In Argentina, the economic arguments for employing improved pastures had gained strength after the late 1880's with the demands of the European, particularly the British, markets for higher-grade animals.

While ranchers in Buenos Aires Province realized the need to replace the native grasses with planted varieties, most claimed that they could not afford to seed alfalfa in their own right. However, a mechanism for pasture improvement was at hand. Ranchers subdivided their estancias, encouraging a sharecropping system of three-year contracts. Tenant farmers cultivated wheat but left the land sown to alfalfa in the final year of their contracts.[148] Southern European immigrants streamed in the thousands onto the wheat frontier of the pampas to work as sharecroppers and harvest laborers. As Scobie graphically described, wheat farming became a key to ranch intensification in Argentina. In Rio Grande do Sul, by comparison, there was a true pioneer fringe in the Serra. Even during the later years of this study, Italians were still gaining access to their own land in northern Rio Grande do Sul; they had no motive for becoming tenant farmers in the Campanha.

It is much harder to understand why the Campanha ranchers did not do more to protect the better of the native grasses for grazing, such as "flechilha" (*Stipa neesiana*) and trefoil (*trifoliu*). These valuable species, under continual attack from tooth and hoof, regenerated if fenced off. Regeneration began with patches of range, not with leagues; it required more initiative than expense. Assis Brasil saw Brazilian ranchlands during the 1890's as being far behind the Plata in this respect for reasons he identified in lofty Victorian phrasing as "routine, indolence, and moral cowardice."[149] Assis Brasil had a point; however, pasture regeneration required a greater degree of careful property management than most ranchers were ready to undertake.[150]

The economic and cultural barriers to the use of planted grasses in the Campanha remained strong well into the twentieth century. Shortly after World War I, a Riograndense agronomist, Pedro Pereira de Souza, summarized the local attitudes well. Whenever anybody talked of pasture improvement, the ranchers would put their hands on their heads in stupefaction. Pasture improvement involved dividing large properties into paddocks and, worse, plowing, grading, fertilizing, and seeding such paddocks. Worse still, it meant seeding forage crops with names "exquisite, unfamiliar, and exotic."[151] Ranchers mistrusted the costs associated with the grasses. Their nativist language reflected this, but it also served to obscure their practical material inability to deliver these improvements.

The Campanha could not follow the example of Buenos Aires Province in the widespread cultivation of alfalfa for two major reasons. First, the physical environment of the Campanha was less suitable for alfalfa;

many of its soils, in fact, were unsuitable. In addition, the continued existence of an open frontier in the Serra diverted potential immigrant labor away from the Campanha ranches. A lack of what contemporary observers termed "suitable" labor rendered the conversion of natural pasture to alfalfa paddocks difficult and usually prohibitively expensive. Unlike their colleagues in Buenos Aires Province, Campanha ranchers rarely had anybody to shoulder the risks of pasture improvement for them. As a result, the seeding of grasses was left mainly to enthusiasts in the Campanha.

Veterinary Improvements

By World War I, both federal and state governments were extending limited veterinary assistance to the ranchers, organizing vaccinations, for example, for tuberculosis-infected animals and against *tristeza* (Texas fever). Vaccination reports of government veterinary missions into the interior of the state became a new theme for newspaper reporting at Porto Alegre, with rancher responses ranging from publicly stated gratitude to complete indifference.[152] Livestock diseases posed major problems for Riograndense breed improvers; even on carefully managed estâncias, such as those belonging to Assis Brasil, mortality rates of imported animals were high.[153] In addition to veterinary help, both levels of government provided subsidies for the construction of concrete tick-baths (*banheiros carrapaticidas*). These were very expensive to build, partly because they often entailed drilling deep wells to ensure a water supply.[154] The baths played a vital role where ranchers bred cattle from European breeds on tick-infested pastures. Ranchers knew that these animals needed to be dipped in vermin-killing liquid, or they would likely succumb to tristeza. Ticks were also a problem for sheep. Smaller concrete baths were constructed to dip them against *sarna* (scab). Tick-baths are a secondary index of modernization in that they accompanied the diffusion of improved breeding.

In 1914, there were 86 tick-baths for cattle in the whole of Rio Grande do Sul, nearly half of them in Bagé and Itaqui.[155] During the First World War, the number of tick-baths increased steeply, in step with the greater emphasis on improved cattle breeding. Figure 5.6 shows the distribution of the 374 tick-baths present in 1918; it uses an index of concentration (IC), the proportion of tick-baths present in a county divided by that county's proportion of total state area. Figure 5.6 provides a very crude indicator of the regionalization of improved

Fig. 5.5. Bathing the cattle, c. 1916. Photograph by Robles. From *A Estância* (Apr. 1916): 108.

cattle breeding in Rio Grande do Sul at that time. The predominance of the tick-baths in the Campanha counties is striking. Baths showed their highest degree of concentration in the southeastern fattening grounds around the leading sites of slaughter. Thus there is a clear zone of seven adjacent counties running from Bagé to Pelotas where indices surpassed 2.0; within that zone, the indices were much higher along the Uruguayan frontier. A second zone of concentration covered the southwestern part of the state from Livramento (IC 2.7) through Alegrete (IC 2.3) to Itaqui (IC 2.7). Some high values around Porto Alegre (notably at Gravataí, with an IC of 5.0) are probably more reflective of the availability of money from the state capital than of ranching potential. Taken together, the counties where the baths were concentrated were likely those where most of Rio Grande do Sul's cattle improved from European breeds were to be found in 1918.

The distributions examined in this section reflect the character of the data; they are valuable but crude. Any search for detailed spatial congruence between all the patterns is likely to founder. For example, Bagé and Dom Pedrito, adjacent Campanha counties with similar physical potentials for ranching, show inverse levels of membership of the Breed-

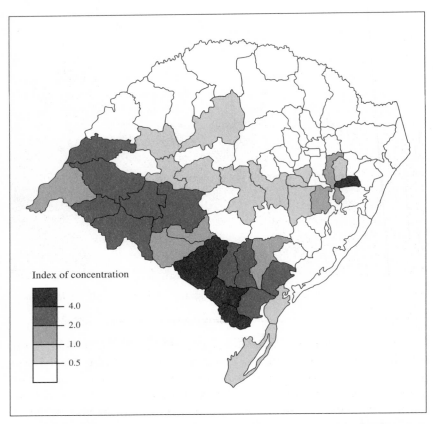

Fig. 5.6. The distribution of tick-baths in Rio Grande do Sul, 1918. Data from State of RGS, *Relatório de Estatística, 1918*, p. 70.

ers' Union and numbers of tick-baths.[156] Membership of agricultural societies clearly did not always translate into results. If we concentrate on the material evidence for ranch innovation, some broad relationships do emerge. Most of the tick-baths in 1918 were located in counties where ranchers had still not adopted the Zebu (not susceptible to the cattle tick) in 1914 (Figure 5.1), implying that they were concentrating on European breeds. In addition, the distribution of the cattle baths thinned in the southwestern and southeastern extremities of Rio Grande do Sul, the areas paying most attention to sheep (Figure 5.4). Bearing in mind that cattle were still the leading concern of Riograndense ranchers in 1920, the county distribution of tick-baths (Figure 5.6) provides the best approximation of the regionalization of improved ranching.

The patterns of innovation on the ranches of the Campanha bore only limited resemblance to those in the core of the southern South American ranching region. On the humid pampas of Buenos Aires Province, wheat cultivation provided a mechanism for ranch intensification as tenant farmers left the land sown in forage grasses. In the Campanha, there was no hope of replicating Buenos Aires in the conversion of natural to planted pastures. Working on natural pastures, and without the early pull of the European markets, Campanha ranchers accomplished the intensification of their land use in a different, and slower, way. After 1880, they gradually adopted fencing and crossbreeding of cattle to meet the uncertain signals of a growing salt-beef industry. By World War I, these gradual improvements on the estâncias had laid the basis for an intensified effort at selective breeding and more subdivision of the ranches into paddocks, both of which were key to the provision of raw material suitable for the frigoríficos.

CHAPTER SIX

From Charqueada to Frigorífico

THE KEY DEFICIENCY of charque was tightly circumscribed markets. While the taste of salt-beef perhaps improved with aging, few ate it by choice. Slaves in other parts of Brazil, and to a lesser extent Cuba, were the consumers of most of Rio Grande do Sul's charque. A key element in the modernization of southern South American ranching was the search for broader and more lucrative markets. This required major changes in the products of ranching. As with the indicators of modernization already examined, response to market stimuli varied across the southern South American ranching region.

The commercial crisis of the early 1860's provided the first serious incentive to search for new markets for southern South American meat products. As the prices for salt-beef fell, the scope for experimentation with more distant markets increased. Early initiatives focused first on possible new markets for the best of the salt-beef. In the early 1860's, John D. Jackson, an important Anglo-Uruguayan rancher, had plans to supply Britain, continental Europe, and China with salt-beef, using his family connections with Rathbones of Liverpool (a mercantile firm with worldwide connections) to market the product.[1] Nineteenth-century science also turned to the chemical preservation of "extractum carnis" (meat-extract). The establishment in the 1860's of the Liebig Extract of Meat Company's famous plant at Fray Bentos in Uruguay was a landmark innovation in the development of new products that could expand the markets for cattle. The company was soon selling large quantities of meat-extract to the armies and the institutional poor of Europe. For existing salt-beef interests, like those in Rio Grande do Sul, the challenge was to escape the limited markets of Brazil and Cuba, while the reform-minded stressed the importance of improving the product first.

In the 1860's, Rio Grande do Sul experimented with selling charque in Europe. Although unsuccessful, these efforts had a lengthy legacy; a half-century later there were still optimists in South America convinced that they could find new markets for salt-beef.[2] In the later 1880's, two important innovations came to Rio Grande do Sul that had major implications for the salt-beef industry—the abolition of slavery and the arrival of railways. Both of these greatly weakened the century-long predominance of the Litoral in Riograndense cattle slaughter. The last decades of the nineteenth century saw three major changes in the treatment of southern South American ranch products; the transformation of the salt-beef plants to operation on an industrial scale; the export of animals on the hoof to Britain; and, most importantly, the introduction of refrigeration technology in the form of frigoríficos, the long-term answer to the modernization of ranch products for worldwide markets.

During the last third of the nineteenth century, while shipping frozen meat over long distances was still experimental, South American ranchers made serious efforts to send live animals to Europe. This became a large business, especially with Britain. In some of the last years of the nineteenth century, Argentina remitted up to 100,000 bullocks directly across the Atlantic; but the export of live animals had little significance for Rio Grande do Sul. Work on dredging the bar at Rio Grande, necessary to permit access by ships large enough to carry livestock across the Atlantic, did not begin on a serious scale until 1906, when the traffic was already coming to a close. Even the Argentinean traffic in live animals did not last; its economic basis was fragile, especially in terms of transport costs, and the British were quick to close their ports whenever they saw any prospect of foot-and-mouth disease in Argentina.[3] Nevertheless, the export trade in live animals was important; it provided Argentina's ranchers with a new market during the long phase of experimentation needed to make the export of refrigerated beef commercially viable, and it provided the impetus to begin the refinement of their raw material.

The long-term answer to the modernization of ranch products lay with refrigeration. Various attempts were made to invent an efficient refrigeration system starting in the 1830's, if not earlier. Much of the early serious development of refrigeration took place in the United States, where meat producers on the western frontiers had the strong incentive of a distant but large domestic market in the growing cities of the eastern seaboard. As early as the 1860's, American railways used iceboxes to move large quantities of meat from the Midwest. By the 1880's, railway cars using mechanical refrigeration were common in the United States. In the forefront of technical improvements, Chicago-based Swift and Armour were the two largest meat-packers in the world by the 1890's.[4]

Europeans were particularly interested in extending refrigeration to ships, so that they could import supplies of meat. Charles Tellier's invention of a machine using sal ammoniac in 1872 made an international refrigerated-meat business possible. Even this was far from being an automatic success, but by the mid-1880's, after trials in Australia and Argentina, it was proving effective for the transport of mutton from Argentina. The first direct shipment of frozen mutton from Buenos Aires to Liverpool occurred in 1885.[5] By the late nineteenth century, techniques of refrigeration had become sophisticated enough that meat could be shipped to Europe either deep-frozen (kept at $-10°C$) or chilled ($-1°C$).[6] This led to a new segmentation of the market. While frozen beef was "effectively imperishable," freezing involved some loss of taste and as a result made the product less attractive to the consumer than chilled beef.[7] Chilled beef was a much better product but required a more refined raw material than frozen beef. Market demand for chilled beef sent stronger signals for estancia transformation than did the demand for other cattle products. Chilled beef lasts for only about forty days from the time of slaughter, not long considering the distance (approximately 11,000 kilometers) between the sites of production along the Plata estuary and the retail butchers of Britain's high streets. These constraints provided much of the rationale for the early integrated corporations that controlled every stage in the chilled-beef trade, from production to sale overseas.

The international trade in frozen meat had massive implications for South American ranching. By 1920, Argentina alone was supplying approximately 80 percent of Britain's beef imports. However, like those aspects of the modernization of ranching already discussed, frigoríficos and the important new technology they represented were slow to appear in Rio Grande do Sul.

Broadening the Markets for Charque

The commercial depression of the early 1860's made merchants throughout southern South America think hard about how to conquer new markets for salt-beef. Brazil's fiscal structure limited the feasibility of exporting charque to new markets, even with the low salt-beef prices of this period. At the time, an arroba of charque was selling for as little as 500 to 600 réis; the salt used to produce that arroba of meat cost an estimated 200 réis and the transport from Pelotas to Rio Grande, the port of export, another 80.[8]

To bring about a widening of trade, one of the leading British merchants at Rio Grande, John Gardner, asked the provincial assembly of Rio Grande do Sul to foster new efforts at marketing local products. The provincial assembly was also urged by its Liberal members to petition the imperial government, in order to see whether it was possible to remove the burdensome 7 percent tax placed on charque exported beyond the Empire.

Concern about the capacity of Rio Grande do Sul to sell its charque in Europe also surfaced during the assembly's debate about new markets. For example, deputy Miguel Pereira de Oliveira Meirelles argued that if Rio Grande do Sul expected to sell its salt-beef in Europe, the product would have to be of the best quality, not that which sold for 1$000 or less per arroba; he questioned charque's palatability as an item of normal consumption for people other than slaves.[9] While Meirelles was dubious that the food habits of the English poor could be changed at a stroke, deputy Nascimento believed that poverty would induce them to eat salt-beef if the price were low enough.[10] Others debated where any provincial help should be directed: to charqueadores, who had some capacity to set prices; to ranchers, who had no outlet for their products other than the coastal slaughtering plants and had to accept the prices offered; or to the abolition of slavery and the move to free labor.

The result of this lengthy debate was that in 1862, partly due to Gardner's efforts, the province decided to pay an award of eight contos (around 880 pounds sterling) to whoever managed to introduce "in whatever part of Europe, a load of salted meat of whatever kind, in perfect state, never less than 8,000 arrobas [120 tons] and manufactured in this province."[11] A year later, Esperidião Barros Pimentel, the provincial president, informed the legislative assembly that no one had taken up the challenge; in fact, no companies had yet appeared in Rio Grande do Sul for the exportation of preserved meat. Pimentel, however, had been directly involved in bringing to Porto Alegre samples of meat preserved by different methods in Buenos Aires. These methods involved packing dehydrated meat into tins under vacuum and injecting the tins with preservatives.[12] The meat brought from Buenos Aires had been opened at the Commercial Association in the presence of the inspector of public health. Even if some of this meat was rather blue in places, similar to what might be found "in charque burnt by the salt," it nevertheless represented progress. The president urged the local charqueadores to persuade themselves of the "urgent necessity" of trying similar experiments.[13]

Strong presidential interest in finding new markets for salt-beef was largely a result of the weak provincial treasury. Ranching in Rio Grande do Sul displayed some of the weaknesses of a monoculture. When the prices for cattle products declined, as during the American Civil War, there was no other economic activity to fill the provincial coffers. The major problems of the 1860's were the ruinous competition from the Plata and the lack of a secure market. Charque was produced in greater quantity, but the value of that production was decreasing. Not much progress was being made in matters of meat conservation. All the systems tried in Rio Grande do Sul involved the use of hermetically sealed jars and proved unsuccessful. Meanwhile, the Río de la Plata countries were reaching both the Brazilian and the European markets with their higher-quality salt-beef.[14] Much of the contemporary explanation for why the Río de la Plata countries turned out better salt-beef than Rio Grande do Sul concentrated on the extensive use of free labor in the saladeros of the south.

Worries at the presidential level were no guarantee of action in Rio Grande do Sul. If trade in charque was unstable, it remained a large business. And it was a business that had seen growth, although irregular, since mid-century. For example, during the 1860's annual exports of salt-beef from Rio Grande averaged approximately 31,000 tons, and in the 1870's 26,000 tons.[15] Except when trade was depressed, as during the early 1860's, the issue of abolishing slavery in the slaughtering plants masked the urgency of developing new markets for ranch products during the final decades of the Empire.

Changes in Charqueada Location Patterns (1889–1920)

During the last decades of the nineteenth century, the Riograndense charqueadas saw important changes in their location and character. These changes largely reflected the shift from slave to wage labor after abolition in 1888 and the arrival of railways in the Campanha.

ABOLITION OF SLAVERY

It has often been noted that slavery was not as deeply entrenched in Rio Grande do Sul as in the plantation zones of Brazil. There is much justification for this claim. For example, slavery played little part in the evolving agriculture of the colonial zones; after 1854 the European colonists were not permitted to use slave labor on their farms.[16] On the other hand, slavery was important to the pastoral economy. Rio Grande

TABLE 6.1

Slave Population in Selected Counties of Rio Grande do Sul

County	1884	1885	1887
Alegrete	1,200	30	0
Bagé	2,435	1,001	82
São Gabriel	1,636	137	0
Pelotas	6,526	2,831	338

SOURCE: Data from Bakos, *RS: escravismo & abolição*, pp. 22–24.
NOTE: Total slave populations for the province as a whole were 62,231 in 1884, 22,042 in 1885, and 7,901 in 1887.

do Sul was very different in this respect from its southern neighbors during the second half of the nineteenth century. The persistence of slavery into the 1880's cast a somber economic shadow over Brazil's southern frontier.

The limited statistics available imply that the slave population in Rio Grande do Sul peaked in the mid-1870's at almost 100,000.[17] The Law of Free Birth of 1871 and other measures aimed at gradual abolition brought a general decline after that time. However, in 1884 there were still over 60,000 slaves in Rio Grande do Sul, placing it sixth in the list of Brazilian provinces with the largest slave populations.

Slaves were still present on ranches in the early 1880's. For example, Francisco Martins da Cruz Jobim, who had registered 34 slaves at São Gabriel in 1872, had 18 still present at his ranch, Salso, in 1880.[18] Despite the continuing use of slave labor on some estâncias, impending abolition was not a major issue for ranchers. Their adoption of wire fencing was already leading to a reduction in the number of workers on the ranches, whether slave or free. In 1883, Rio Grande do Sul established a general tax on slaves of 4$000 per capita; the subsequent tripling of this tax had a decisive effect on abolition in Campanha counties (Table 6.1). Ranchers directed their worries to the potential threat to property that ex-slaves might pose after they had been freed.[19]

In contrast, slavery was still the key to charqueada operation at the opening of the 1880's. According to contemporary descriptions, established Pelotas slaughterhouses employed 60–90 slaves each at that time, as well as some free labor for the more specialized tasks, such as boiling grease.[20] Although the abolition movement had a large impact on Rio Grande do Sul in general after 1884, some plants continued to maintain considerable concentrations of slaves.[21]

Many of the limitations of producing salt-beef using slave as opposed to wage labor had been identified in the early nineteenth century> With wage labor, the employer could adjust the structure of the labor force (according to ages and skills, for example) at will. Capital previously "buried" in slaves could be diverted toward material, notably technical, improvements. Since salt-beef production was a seasonal activity, charqueadas underutilized their slaves during the slack season. And, as Gonçalves Chaves had noted in the 1820's, the obvious human response of those reduced to slavery was to try to work as little, and consume as much, as possible.[22]

Around 1880, the increasingly successful competition mounted by the Río de la Plata wage-labor saladeros became a specific concern of the Empire. The imperial government asked Louis Couty, a Frenchman who had arrived in Rio de Janeiro in 1878 to give a course in industrial biology, to make a comparative study of charqueadas in Brazil and their counterparts in Argentina and Uruguay.[23] The leading theme of his 1880 report was the greater productivity of the paid labor force in the Río de la Plata saladeros. By 1880, specialized workers for the more important tasks were an established feature of the Rioplatense plants. While no single slaughterhouse in Pelotas was killing more than 20,000 head of cattle in 1880, some of their Uruguayan competitors were already much larger; they had achieved important economies of scale by using motivated, specialized wage labor.

Pelotas families made some considerable fortunes from their charqueadas, but they had difficulty making the mercantile-industrial transition. By the 1880's, retention of slavery in the Pelotas plants had placed southern Brazil far behind the Plata in adopting new technology. When British interests were undertaking practical experiments in freezing sheep in the early 1880's (and not only in Argentina), the Pelotas industrialists were still following their established pattern of producing salt-beef using predominantly slave labor.[24] It took the abolition of slavery in 1888 and the construction of railways to the Campanha to shatter the near-monopoly of the Pelotas mercantile elite in the Riograndense cattle-slaughtering business.

TRANSPORT

With little promotion of their economic value by either ranchers or charqueadores, railways came late to the Campanha. The construction of a railway from Pelotas to the leading towns of the Campanha had been mooted at the beginning of the Paraguayan War, but the first line seriously considered was intended to connect the coal mines at Candiota

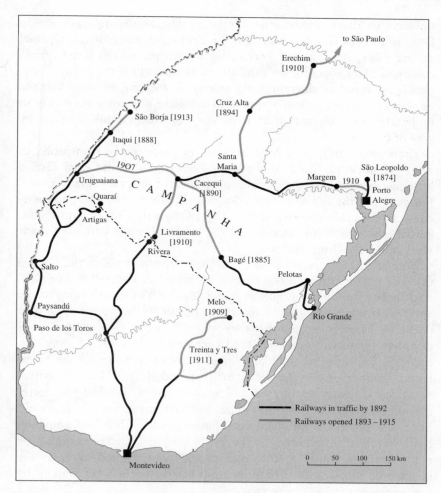

Fig. 6.1. Railways in and toward the Campanha, 1915. The dates in brackets indicate when the railway reached particular locations. Data from Barrán and Nahum, *HRUM*, 7: 129, and Roche, *Colonização alemã e o Rio Grande do Sul*, 1: 64.

with the coast (the forerunner of the railway that commenced traffic in 1885 between Rio Grande and Bagé). The Paraguayan War had exposed the weaknesses in communications along Brazil's southern frontier, but the argument for building railways as strategic links became a matter for political debate. Although politicians may have discussed railways to the Campanha as a matter of strategic importance, Rio Grande do Sul's first line was built for clear economic motives. Opened during 1874, it connected Porto Alegre with the established German colony of São Leopoldo, with the aim of siphoning off the growing surplus agricultural production of the colony. It was only in 1878 that the imperial government finally agreed to subsidize any company ready to undertake the construction of a railway from Rio Grande to Bagé by way of Pelotas.

There was barely a skeletal network in place in the Campanha in 1892 (Figure 6.1). Much of the limited trackage that did exist carried little significance for the economic integration of Riograndense territory. For example, the Brazilian Great Southern Railway, a British-built, narrow-gauge line which reached Itaqui by 1888 (early by Campanha standards), was simply an extension of the standard-gauge Uruguayan network along the Uruguay River.

Uruguayans, on the other hand, in their drive to develop the volume of trade passing through the port of Montevideo, had followed an explicit policy during the 1880's of building railways towards some of the key border zones, in order to secure the transit traffic from Rio Grande do Sul. It has been argued that they gave this policy even greater attention than the matter of serving the Uruguayan rural economy more efficiently.[25] By 1891–92, standard-gauge railways had tapped the Campanha frontier at San Eugenio (Artigas) and Rivera. Their Brazilian equivalents, lightly engineered and narrow gauge, reached the corresponding Campanha towns of Quaraí and Livramento only in 1939 and 1910. The provision of railways was late not only within the Campanha itself. While the first part of the Porto Alegre–Uruguaiana link was ready in 1883, those in the state capital who wanted to cross the breadth of Rio Grande do Sul by train had to begin their journey by river-steamer until as late as 1910.

Once in place, the railways of Rio Grande do Sul exhibited many deficiencies of operation, summarized in 1901 with admirable brevity by one rancher-spokesman: they were foreign-owned, linked distant nuclei of sparse population, and provided services only at high cost.[26] The impact of railways as a delivery system long remained limited for major aspects of the ranching economy. As late as 1902, the state government of Rio Grande do Sul was still negotiating the carriage of live animals by rail (the holders of the concessions for the federal lines required permission from Rio de Janeiro to borrow the money needed to build specialized rolling stock).[27] Nevertheless, railways provided the technical means for the dispersion of the charqueadas into the breeding zones of the Campanha itself, especially those close to the frontier with Uruguay. Despite all of their operational weaknesses, by the 1890's railways were used to transport salt-beef.

During the early decades of the Old Republic, the location pattern of the charqueadas diversified (Figure 6.2 shows the pattern of all the slaughtering plants active in 1920). Pelotas remained an important point of slaughter, but abolition had weakened the hold of the traditional Pelotas families. After abolition, the processors had much more flexibility in their sources of labor recruitment. With the emergence of a regional railway system, plants were no longer constrained to the Litoral by transport considerations, and they began to appear in the Campanha frontier towns, closer to the areas where the cattle were fattened.

The establishment of charqueadas in the Campanha frontiers was also a reflection of new investment attracted by tariffs. Riograndense salt-beef processors won a measure of protection for their industry after 1886, a measure designed to help them compete with the Plata during their transformation from a slave- to a wage-labor system. Under the more decentralized administration of the Old Republic, this tariff protection intensified, especially after the civil war of 1893–95.[28]

Crossley and Greenhill have interpreted Brazilian protection of the domestic meat industry under the Old Republic as interrupting the "natural flow" of cattle southwards.[29] In Argentina, cattle were bred in the provinces of Corrientes and Entre Ríos but fattened further south. On the east bank of the Uruguay River, the political economy of the international boundary between Brazil and Uruguay conditioned the flow of cattle. However, the pattern of drawing cattle from northern Uruguay into Rio Grande do Sul for slaughter was by no means a new development of the 1890's; it was as old as the century.

Tariff protection meant that the Rio Grande do Sul slaughterhouses could pay higher prices for fat cattle, and this had repercussions on the structural form of the meat trade throughout the broader pastoral region. By paying good prices, Campanha charqueadas continued to attract cattle from northern Uruguay brought over the international boundary for slaughter in Brazil. Montevideo, however, served as the entrepôt for much of the salt-beef produced in Rio Grande do Sul and destined for sale in Brazil's major cities. This reflected Uruguay's infrastructural advantages over Rio Grande do Sul, advantages in both its leading port and its railway system.

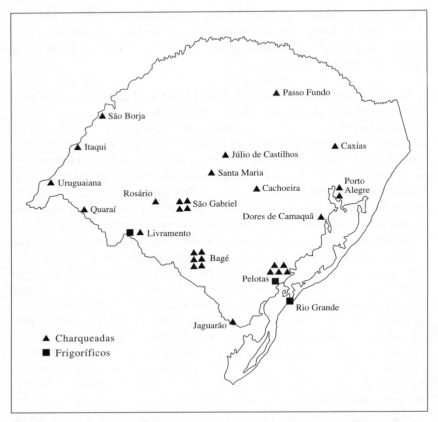

Fig. 6.2. Locations of charqueadas and frigoríficos in Rio Grande do Sul, 1920. Data from *Revista do Arquivo Público do Rio Grande do Sul* 8 (1922): 240–45.

The tariff wall attracted extensive new investment from the south in salt-beef plants along the Campanha frontiers. In addition, as the sphere of influence of the Rioplatense frigoríficos (first established during the early 1880's) extended further into the hinterland, their ability to pay the highest prices for cattle forced changes on the longer-established plants producing cheaper pastoral products, such as salt-beef and meat-extract. These producers, unable to compete in the Río de la Plata, were attracted to pools of cattle in ranching zones farther removed from the competition of the frigoríficos. The Campanha was one such zone. It had the additional advantage of the Uruguayan railway system, which enabled Rioplatense processing interests to continue to market Riograndense salt-beef using their existing commercial connections in the estuary. This was the rationale, for example, behind George Clark Dickinson's establishment of a large charqueada at Itaqui as late as 1910. By 1913, Uruguayan interests controlled ten of the eleven charqueadas located along the Campanha's southern frontier.[30]

These charqueadas established after the 1880's in the Campanha frontiers bore only limited resemblance to the Pelotas slaughtering plants of the mid-nineteenth century. The new plants exhibited a far greater degree of industrial organization than their earlier counterparts on the plain of the São Gonçalo. Some of them, for example, included departments for canning corned beef and ox-tongues. The industrial equipment of the Campanha plants, while limited in comparison with that of the later frigoríficos, made some of them big concerns by Brazilian standards of the time. Riograndense charqueadas occupied ranks 8, 16, 20, 31, and 32 among the 40 largest industrial enterprises (by value of production) in Brazil surveyed by the Centro Industrial do Brasil in 1907.[31]

There were practical limits to the height of the protective tariff wall along the Brazilian frontier. In the early twentieth century, Rio Grande do Sul did not have the production capacity to serve the whole of the Brazilian salt-beef market (it was still rebuilding its herds after the destruction of the civil war of 1893–95), and it could not close out competition entirely. Raising the price of salt-beef through import tariffs was also politically sensitive, since by now its largest market lay with the Brazilian urban poor. Increasing the price of Brazilian charque to satisfy the ranching lobby from the southern frontier also stimulated states like Goiás, Minas Gerais, and Mato Grosso to hasten to improve their contributions to the national pastoral production and increase their market shares.

Under Brazil's Old Republic, charqueadas attracted large-scale investment. But the success of the salt-beef industry in Rio Grande do Sul was largely dependent on the degree of competition from the Río de la Plata. After 1906, when Argentina moved decisively into frozen-meat production and Uruguay began to follow, Rio Grande do Sul worked to take up the resulting slack in the traditional markets for salt-beef, and regained markets from which it had been displaced decades earlier, notably Cuba. Charque remained a highly volatile business, but the quantities produced in Rio Grande do Sul grew with the intensification of the ranches. For example, in the period 1909–13, annual salt-beef exports from Rio Grande do Sul averaged over 60,000 tons, more than three times the quantity at mid-century.[32] Nevertheless, by World War I, ranchers knew that the prospects for sustained economic development lay with frozen meat.

Frigoríficos

Lack of foreign interest and investment is surely the leading factor behind the absence of frigoríficos from Rio Grande do Sul before World War I. Frigoríficos represented the arrival of foreign capital on a large scale in southern South America. The timing of their arrival was linked in each place to a blend of foreign and domestic interests. External interest concentrated first on the core region around Buenos Aires, where the effective demand for the innovation was highest. It came to the peripheries only later. While the early concentration of external interest on the Plata is readily understood, the appeal of Brazil's vast, undeveloped center-west seems less obvious. It attracted foreign investment before Rio Grande do Sul, something that requires explanation. What drew foreign investment toward the Mato Grosso at the expense of Rio Grande do Sul?

In South America, sustained early interest by the great British and American meat-packing companies concentrated on Buenos Aires, the heart of the pampa ranching region. By 1882 British interests had developed a freezing plant there, the River Plate Fresh Meat Company Limited. Competitors were quick to follow. Colin Crossley, a geographer who has studied the organizational form of the River Plate meat trade, has argued that the foundation of the earliest South American freezing plants in the Plate estuary followed spatial logic; these were the best locations to minimize transport costs.[33] A mix of cultural and institutional factors also guided British companies toward the estuary of the Plata. The necessity of breaking into new international markets for meat was a long-standing national issue in Argentina. Investment in packing plants further integrated British mercantile interests already established in such areas as banking, railways, and insurance; for example, a leading backer of the River Plate Fresh Meat Company was George Drabble, chairman of the London and River Plate Bank since 1869 and already active in resource development throughout the Plata for decades.[34]

Technical reasons also favored the Plata. Early efforts in refrigeration concentrated on the smaller carcasses of sheep, in the 1880's still a weak element of the ranching in the Campanha. As far as cattle were concerned, with its history of live-animal shipments to Britain, the Río de la Plata estuary already contained the beginnings of a higher-quality pool of the crossbred livestock that the freezers demanded. These factors gave British entry into the Plata its massive commercial success. That success was not lost on the big Chicago packing interests, and with the new century, American firms moved swiftly into competition. The American packers placed major emphasis on developing a high-grade chilled-beef industry, one that was already important in Argentina by 1914.

During the first decade of the twentieth century, the export of frozen meat became a huge business around the Río de la Plata, in terms of both the volumes handled and the profits made. The technical constraints (the quality of the raw material and the capacity to keep the meat adequately frozen during long ocean voyages) that had dogged the industry in its earlier years were being removed. It was becoming entirely feasible to ship cattle carcasses the long distance to Europe.

The idea of forming a national frigorífico took root in Uruguay in 1902, motivated in part by the huge profits that British freezers in Argentina had recently declared. In 1905, La Frigorífica Uruguaya began production for the export market.[35]

The lapse in southern Brazil between the first discussion of packing plants and their realization was much longer than in the Plata, although experiments there in the 1870's had repercussions in Brazil. In 1876, the imperial government conceded the right to take frozen meats and fruits into the northern sections of Brazil to a lieutenant in the navy.[36] By the 1880's that right had passed to other interests, including Junius Brutus Cassio de Almeida, a Pelotas charqueador. In early 1889, meetings in Rio Grande do Sul discussed the advantages of the frigorífico system, with the regional newspapers providing detailed analysis of the business.[37] Despite the efforts of local experts, the frozen-meat enterprise struggled to find subscriptions within Rio Grande do Sul.

The crucial shortage of local capital was only one of many reasons. In his regional almanac for 1888, Graciano Alves de Azambuja correctly observed that Rio Grande do Sul had no hope of competing in the freezer business until vessels of between 3,000 and 5,000 tons could cross the Rio Grande bar easily.[38] Skeptics raised a variety of arguments against the undertaking; the meat of Rio Grande do Sul was not suitable for freezing since it was too tough; Brazil had a weak entrepreneurial spirit that did not mesh with the true productive capacity of the country.[39] Such comments brought spirited defenses from within Rio Grande do Sul. An Argentinean opinion that the Brazilians or Portuguese were only interested in immediate economic returns was rejected with the telling observation that most of the recent changes in the pastoral industry around the Plata had in fact little to do with people of Iberian descent.[40]

Despite the skeptics, the case for the transformation of Rio Grande do Sul's ranching had become much clearer. Statistics like the following provided eloquent argument: in the eighteen-month period ending in December 1888, around the time of the first successful transport of frozen meat from Pelotas to Rio de Janeiro, Britain imported 939,000 frozen lambs from New Zealand, 908,000 from the Río de la Plata, and 108,000 from Australia.[41]

The government of Rio Grande do Sul was struck by the impact of the establishment of large foreign-owned frigorífico concerns in Argentina, where, by 1906, most beef exports were either frozen or canned.[42] Beginning in 1903, governor Borges de Medeiros courted a British concern, the Brazilian Cold Storage and Development Company, with concessions, but nothing came of the negotiations. While there was still room for expansion in the Plata estuary, the Campanha had limited drawing power for foreign packers. Infrastructure was weak, and the raw material remained of marginal quality for freezing.

Continuing lack of agreement among domestic investors also slowed the development of frigoríficos in Rio Grande do Sul. While some of the farsighted made the case for packing plants, others made strong counterarguments. During the early Old Republic, it was argued that breeds of improved quality were less important than price in the broad domestic segment of the market. Almost all of this was built around cheap meat, a product that only well-managed creole cattle could supply.[43] As late as 1913 there were still deluded optimists in parts of South America who maintained that European markets would accept beef from creole cattle.[44] Others were simply fatalistic about the developments in the Plata. They doubted Rio Grande do Sul's ability to compete: the animals of the Plata were generally of better quality, as were the wintering methods used (a reflection of the greater availability of forage); Argentina and Uruguay enjoyed lower transport costs; and there remained the constraints of the Rio Grande bar.

Aware of the monopolistic tactics that packing plants had adopted in the Río de la Plata, the ranchers belonging to the Breeders' Union were certain that they needed a domestically owned frigorífico in Rio Grande do Sul.[45] Just before the First World War, the subject came under intense discussion, yet results were slow to materialize. This was mainly a reflection of the ranchers' shortage of capital, a weakness again intensified by their lack of consensus. They expended much energy during 1915 in a heated debate about location. Ranchers from Rosário made a strong bid for their district, but other investors believed that a site on water made more sense and promoted Rio Grande. The former site meant refrigerated railway cars; the latter, moving cattle to the point of slaughter by rail.

The Riograndense effort to raise capital and the location debate dragged on into the early war years without concrete results, widening the already visible gap between the condition of pastoral industries in southern Brazil and in the neighboring Plata. Just at this time, a serious and unexpected new competitor entered the field—south-central Brazil. São Paulo, in particular, benefited from a favorable interplay of foreign and domestic interests. Elements of the established coffee-planter elite had been quick to mobilize their wealth (considerably greater than that to be found in the Campanha) toward the establishment of packing plants. These elite families viewed ranching for the export market as a promising means of diversifying land use on their worked-out coffee plantations. Unlike that of Rio Grande do Sul, the state government of São Paulo was quick to subsidize their initiatives in packing plants, a reflection of the planters' political power.[46]

In addition, even before World War I, south-central Brazil attracted major foreign investment in ranching in the form of the Brazil Land, Cattle and Packing Company. This was one of the 38 subsidiary companies of the Brazil Railway, a giant trust of railway and resource development companies headed by Percival Farquhar, an American entrepreneur.[47] Farquhar's land and cattle company controlled vast rangelands in Mato Grosso and Goiás, and it bought and leased other properties for fattening cattle within easy reach of São Paulo. Cost rather than the quality of the raw material was the principal factor in this company's locational rationale. It was much less expensive to assemble millions of hectares for a company enclave in Mato Grosso than in Rio Grande do Sul. In addition, the Mato Grosso and Goiás location linked the land company to the expanding railway network of São Paulo (much of it also controlled by the Farquhar group) and to the port of Santos, improved as far back as the 1870's. The superior infrastructure of São Paulo before World War I again reflected the power of the coffee economy.

Although a recent study sees the Farquhar initiatives as important in their own right for their impact on Brazilian ranching methods, the concrete results of this huge capital investment are debatable.[48] The clearest impact of Farquhar's ranching schemes lay in alerting other American companies to the economic potential of Brazil, and notably the greater São Paulo region. Foreign capital probably underestimated the potential of Rio Grande do Sul for freezer plants before World War I, a demonstration of Hirschman's thesis that "private investors consistently overestimate the profitability of investments at the center relative to the periphery."[49] It is instructive to compare Theodore Roosevelt's boosterist impressions, gained from a late-1913 visit to one of the Farquhar ranches in northern Paraná, with the reasoned skepticism expressed less than a year later by Assis Brasil, a leading authority on Brazilian ranching.[50] Assis Brasil knew that imported Shorthorns on poor pastures made no economic sense. While respectful of the energy and enterprise of American initiative, he considered the investment in purebred livestock in south-central Brazil a wasted gesture when it could have achieved much in Rio Grande do Sul. The force of this argument that a subtropical region was much more hazardous for the acclimatization of European breeds than the more temperate Campanha was abundantly revealed during the 1920's.[51]

In January 1915 Brazil made its first shipment of frozen meat from Santos to Europe. While the cargo was modest, comprising only 10 tons of frozen beef destined for Geneva and approximately 100 tons for London, São Paulo entered the refrigerated-meat industry earlier than Rio Grande do Sul, despite the generally inferior quality of its herds.[52]

The idea that the future of ranching was bound up with the refrigerated-meat industry took hold more firmly in southern Brazil with the outbreak of World War I. The special economic circumstances of the war presented new opportunities for the establishment of frigoríficos in Brazil, in part since the market for frozen meat was temporarily less demanding about quality.[53] In addition, the United States and Argentina, rival producers of beef, were making significant efforts to intensify their agricultural systems away from ranching. And the United States no longer had meat to export on account of rising internal demand.[54]

There were concerns that prices for meat would fall at the end of the war, but many were optimistic about postwar markets, on the assumption that the effort of war would greatly disrupt the herds of the belligerent countries. Among the optimists was Charles Perkins, co-receiver of Percival Farquhar's Brazil Railway, which had failed in 1915. Perkins felt that meat prices would not return to their low prewar levels. He had spent the early part of 1916 examining the economic conditions of southern Brazil and was confident that Brazil had the basis to build up a first-class cattle industry.[55]

In Rio Grande do Sul, local support for refrigerated-meat plants strengthened as market conditions displayed major shifts brought about by the abnormal conditions of the war. For example, at the beginning of the international conflict, a universal rise in the value of meat increased prices beyond what many Brazilians could pay, hence restricting the markets for charque. In December 1914, at the start of the new killing season in Rio Grande do Sul, there were still around 10,000 tons of unsold charque held in stock in Rio Grande do Sul. Even so, the price in south-central Brazil was about a quarter of that paid for poor-quality frozen beef in Europe. In 1915, the state government of Rio Grande do Sul added extensive tax advantages to existing concessions as a further incentive to attract foreign frigorífico concerns. They brought no results and were extended in the following year.

Along with the growing euphoria of the wartime markets, these incentives of 1916 achieved concrete results. By the end of World War I, three large American companies had established themselves in Rio Grande do Sul. This was part of a pattern of expansion of their already massive investments in the Plata, where they had quickly gained effective control of the industry. For example, the Frigorífico Montevideo, established in 1912 with American capital, was by 1914 killing more cattle than all of the remainder of the Uruguayan slaughtering establishments.[56] The pattern of expansion of frigoríficos is yet another example of how a modernizing frontier diffused through Uruguay to Rio Grande do Sul.

By 1920 two of these American-owned frigoríficos were working in Rio Grande do Sul, at Livramento and Rio Grande. Armour began operations at Rio Grande in 1917 but quickly moved to Livramento, where it bought a major charqueada and spent U.S.$7,500,000 transforming this into a frigorífico.[57] The management provided two motives for the sudden move to Livramento: proximity to the cattle the plant was buying for processing, and the railway link with Montevideo. Armour did not export frozen meat from Rio Grande do Sul until 1920. When it began exporting meat from its frigorífico at Livramento, a standard-gauge spur linked its plant with the nearby Rivera terminus of the Central Uruguayan Railway, which provided direct transport to freezing chambers in a large warehouse at the port of Montevideo.[58] For similar reasons, Wilson was also laying the groundwork for a small fri-

Fig. 6.3. The Frigorífico Anglo de Pelotas in the 1920's. From Brazil, *Recenseamento de 1920*, 5(2), opposite p. 504.

gorífico in Livramento by 1919. Even Swift, which built its frigorífico at Rio Grande, took the precaution of buying a large charqueada at Rosário in the Campanha, holding this in contingency. Swift was freezing meat at Rio Grande by 1919.[59]

As the American companies were taking hold, domestic interests finally built a small frigorífico in Pelotas after years of debate and struggle to raise capital. Commencing slaughter in 1920, this did not survive the postwar recession. By 1921, British interests had bought out the plant, expanding and reorganizing it in 1924 as the Frigorífico Anglo de Pelotas.

Just as Uruguay had entered the frozen-beef trade abruptly on the eve of World War I, so did Rio Grande do Sul after 1919. By 1921, not only had Rio Grande do Sul outstripped São Paulo as a frozen-meat producer, but its frozen-meat exports of 32,548 tons nearly matched those of charque.[60] However, while the freezer technology was in place, growth in the trade could not be sustained. The frigoríficos so recently established in Rio Grande do Sul greatly reduced the scale of their freezing operations during the early 1920's.

The aftermath of the world war saw abrupt changes to the pattern of the meat trade, changes that occasioned economic distress throughout southern South America. In part, these changes reflected the general contraction of the overseas market after the extraordinary conditions of the war. Stockpiles of meat to feed extensive armies were no longer needed. Relaxation of concern for the quality of raw material proved to have been temporary. After World War I, chilled beef regained its market and provided the major growth in the international trade of beef. Chilled beef allowed no room for equivocation over the state of breeding or management technique; it required high-grade, pedigree livestock fattened on nutritious planted pastures. While crossbreeding had advanced greatly in the Campanha during the First World War, around 1920 the cattle there fell between the extremes of quality. They were still not ready to compete with areas like Buenos Aires or southern Uruguay in producing meat for the chiller trade. Nor, on the other hand, were they as cheap as the still predominantly unimproved creole cattle of the remoter fringes of the southern South American ranching region (the Mato Grosso and Paraguay), bred largely for processing into extract or corned beef. As Rio Grande do Sul entered the 1920's, charque remained the Campanha's leading meat product, much to the frustration of local ranchers, who had begun to make costly investments in the future but had yet to reap the benefits on a sustained basis, thus being left halfway up Parnassus.

Innovations invariably came first to Buenos Aires Province and southern Uruguay, the core of the South American ranching region. The broad pattern of diffusion into Rio Grande do Sul was one of lags, lags that varied greatly in both time and degree. For example, the momentum of the northward advance of fencing spilled over the Campanha frontier within a few years at most. Other innovations trailed by decades: the first agricultural societies in Rio Grande do Sul came a full half-century later than those in the Río de la Plata. Finally, some innovations with considerable potential to transform both rural economy and culture, such as the mid-century wool boom, had barely any appreciable effect on ranching in the Campanha. The pattern and pace of modernization were uneven but not haphazard. Each innovation and each delay happened for a reason. In discussing the indexes of modernization in this chapter and the previous two, this study has emphasized the key factors specific to each case. The next chapter treats the broader issues that affected the economic development of the Campanha as a whole.

CHAPTER SEVEN

The Slow Path of Modernization

FOR THOSE WHO lived the period 1850–1920, the gap between core and periphery in the southern South American ranching economies probably appeared widest during the early 1880's. These years marked the beginning of broad-scale changes in southern South American ranching, epitomized by the establishment of the frozen-meat industry and by grain exports from the estuary of the Plata. As Argentina's first packing plants froze sheep for the British market, Riograndenses had still to confront the consequences (including possible economic dislocation) of slavery's abolition, which had taken place around thirty years earlier in the Rioplatense republics. Enthusiasts for change in the Campanha followed the new developments around the Plata estuary with interest, comparing local circumstances unfavorably. Yet, by the 1880's the Campanha showed signs that it was also beginning to modernize; for example, fencing and crossbreeding were both diffusing into the region from Uruguay. Much of the change that took place in the Campanha after the 1880's can be understood through the expansionary momentum of rural transformation in the core, the "spread effect" of modernization emanating from the cities of the Plata estuary.[1] Yet innovations came slowly. This chapter seeks to explain why.

The contemporary literature of improvement laid much of the responsibility for the slow transformation of the Campanha grasslands on the Riograndense ranchers themselves; they were widely identified as poor entrepreneurs. Even today, the idea that Luso-Brazilians are retrograde entrepreneurs in rural matters is still current in the region.[2] This chapter begins with a critical examination of the contemporary case against the Luso-Brazilian as a deficient entrepreneur on the grasslands. While this thesis is seductive, there are other potential explanations for the slowness of Campanha ranchers to adopt new technology. Even in recent decades, ranchers in Uruguay who refused to have anything to do with a large World Bank scheme for pasture improvement, and thus, as Hirschman has observed, were branded "lazy" and "routine-ridden," may have had a point.[3] After the expenditure of nearly U.S.$20 million, the conclusion by the later 1970's was that beef prices were not high enough "to make improved pastures profitable enough to induce farm intensification, as once thought possible."[4] Almost a century earlier, despite the barrage of rhetoric against them, Campanha ranchers may also have had reason to be cautious, waiting for clear rather than obscure economic signals before they made expensive investments in their properties.

A stark clash in values between what the liberal elites wished to see on the estâncias and what was actually taking place accompanied the modernizing frontier. Bradford Burns has argued that the heavy infusion of modernization into Latin American societies during the latter half of the nineteenth century resulted in "what was probably the greatest cultural confrontation in the New World since the early sixteenth century."[5] Civilization against barbarism was a theme played out at a variety of regional scales. Gaúchos of the Campanha form but one example of what Burns would term the "folk." The roots of their pastoral subculture lay in adaptation to the ecological conditions beyond the frontiers of settlement. Gaúcho culture was based on such things as their equestrian habits and traditions, *mate*, and shared frontier language. Burns has argued that "all of this instilled a feeling of unity, loyalty and tradition within the folk, more intuitive than codified."[6] This lifestyle, where "economic decisions took second place to social considerations," repelled most of the elites, who with modernization "became strangers in their own nations."[7] If Campanha ranchers were seen as slow to respond to the vision of the liberal elites, at least some of them were quick to defend an alternative way of life, one that set high store by noneconomic factors.

Even if taken at face value, the comments on rancher conservatism made by local politicians, travelers, and British consuls (all part of the elite) nowhere acknowledge the obstacles faced by those ranchers in Rio Grande do Sul who were keen to innovate, obstacles not shared to the same degree in southern Uruguay or Buenos Aires Province. For example, there were important differences between these regions in matters of property management rights. In Rio Grande do Sul, the state's limited support to landowners in shedding their surplus labor carried important consequences for slowing modernization in the Campanha. The combination of a lack of rural policing and an open international frontier was conducive to threats to property, which in turn weakened landowners' faith in investment.

Infrastructure was a critical issue when neighboring regions were seeking the same markets for their products. It came late to Rio Grande do Sul and remained weaker than in the southern competitors. For example, despite the importance of an effective regional transport system, the groundwork for one had barely been laid before World War I. Early improvements in the quality of Rio Grande do Sul's infrastructure would have compensated the region (at the periphery of temperate southern South America) for its environmental deficiencies and helped it to compete with the Plata.

Why did Rio Grande do Sul miss these development opportunities? The preceding chapters demonstrated that the factors delaying or impeding the diffusion of innovations were sometimes highly specific. However, looking to the Campanha as a whole, several impediments cut across the spectrum of innovation and delayed modernization. Much of the impetus for ranch modernization in southern South America came from the cities. The lack of a major city in Rio Grande do Sul which combined political, trade, and port functions carried major implications for retarding the modernization of southern Brazilian ranching. This absence of urban primacy in Rio Grande do Sul can be linked to differences in the behavior of merchants (important innovators) and to politics.

Retention of Tradition

Complaints about rancher conservatism are deep seated in Rio Grande do Sul. They run back even into the era when the economic pulse of the markets beat faintly.[8] As early as the 1820's, the French botanist Auguste de Saint-Hilaire had identified the existence of a clash of values between the merchants of the coast and the gaúchos of the Campanha. Struck by the imbalance between the wealth of many of the ranchers and their miserable domestic circumstances, he concluded that they must have spent liberally in helping their circles of extended family and friends. Visible wealth was concentrated in the hands of the European merchants in Rio Grande, whom he described as for "the greater part without education and culture" and "inferior to the Americans [the native-born] in spirit and intelligence," rather unimpressive qualifications. The key to understanding this disparity between coast and interior lay with the concept of the future. In Saint-Hilaire's observation, most people in the interior had no sense of this, while the Europeans thought of little else but commerce, profiting from the liberal spending of the native inhabitants.[9]

As southern Brazil was drawn further into the market economy in the course of the nineteenth century, the uneasy harmony of the 1820's between alternative ways of life disappeared. By the 1880's, expressions of dissatisfaction by the liberal elites were legion throughout Brazil. Concern for the future, that key ingredient of any sustained economic development, was far from general in the Campanha. The elites judged most of the ranchers poor entrepreneurs, describing them as "moral cowards," among other favored pejorative labels.

For example, in 1887, Geraldo de Faria Corrêa, the president of the council at São Gabriel, painted a bleak picture of local ranching's condition.[10] He complained of breeding ranches mired in a conservatism highly resistant toward even slight change, and of a monotony of doing the same thing from day to day. He and his councillors at São Gabriel could not help moralizing. Their report bears testimony to the existence of high-Victorian values in Brazil, as can be seen in the observation that "the family ought to merit more from its head than a subsistence." They accused the local estancieiros of doing "a disservice to the public cause" by contenting themselves with routine and ascribed the ultimate cause of the poverty of the county to ranching ("in that wealth, our poverty resides").[11]

Contemporary comment about mindless conservatism ran beyond the local politicians. May Frances, who lived in the Uruguaiana area during 1887–88, wrote a series of letters home to England. They contain every ounce of the stern judgment we might expect from a young middle-class Englishwoman marooned in the camp, yet her comments also place the cultural constraints on rural development in bold relief:

> Sometimes three generations live in the same house, and often a few odd relations live with the family. They spend their days exactly as their forefathers have done for generations, and are as slow as the Chinese about taking in any new ideas; and still continue their old customs, however idiotic they may be, in spite of the way some of their neighbours, a little further south (not Brazilians) are going ahead. They very rarely till the ground, though the soil is most fertile and climate well adapted for growing most of the products now imported; nor do they attempt to stop the degeneration of cattle, sheep, or horses.[12]

Criticism of the ranchers concentrated on the issue of limited crop cultivation. Crops were still generally confined to small corners of the estâncias and were grown largely by the agregado class. At Bagé, the council observed in 1883 that cultivation was limited to "the planting of the most common cereals," grown "through the use of antique and routine processes," so that the county still grew insufficient crops to meet local food demands.[13] And at São Gabriel in the same year, the council claimed that while those ranchers who had made serious efforts to grow crops were doing well, they were few in number.[14] In earlier decades, whenever the ministry of the Empire had questioned the councils of the Campanha about the state of local cultivation, hoping no doubt to hear of positive developments, labor shortage was their favored explanation for lack of action. By the 1880's, however, this was no longer the case. Increasing recourse to wire fencing meant that demand for labor on the ranches was decreasing, except in those rare cases where ranchers had engaged in large-scale crop farming. There was no physical shortage of hands for cultivation.

While they complained that ranchers did not think of the future, the liberal elites were also pessimistic that ranching itself had any future. In a speech made to the provincial assembly in 1885, deputy Villa Nova encouraged more of the ranchers to apply a part of their capital to farming, arguing that ranches were becoming too small and that rising land values made for a poor return on capital, a view repeated almost word for word in the 1887 São Gabriel county report.[15]

Without firm evidence of the economic utility of cultivation, this call for farming was wishful thinking. Such comments assumed that ranchers were a monolithic group and generalized away the constraints on estância management in easy turns of phrase with no serious analysis of the cost economies of the ranchers, whose circumstances surely varied considerably. It is by no means clear whether Campanha ranchers born into families that had owned land and livestock for generations spent much time thinking about the return on their capital; a century onward, modern studies of their descendants incline to view them much more as economic satisficers than as maximizers. A few were wealthy enough to make land-use experiments that produced no short-term return, but most were not; they were short of working capital. Still others had nothing to invest at all; partible inheritance down the generations was slowly squeezing them out of ranching.[16]

During the 1880's, interest in cultivation coming from ranchers themselves was still very limited. They seem to have been caught up in an economic euphoria, brought on by the early phase of the adoption of wire fencing in the Campanha. Thus at São Gabriel in 1883 the councillors wrote of a local perception of success in ranching, linked mainly to improvements in breeding made possible by fences which secured ranchers' property.[17] A similar portrait emerged from Bagé. There the councillors admitted that while the ranching in the county was backward ("em atrazo") in its processes and animal quality, it still displayed "a certain prosperous character."[18] A low return from something understood was preferable to an uncertain return from a new form of land use involving heavy investment costs. The 1887 county report from São Gabriel that complained so bitterly about the excessive conservatism of the ranchers also found an economic basis for their way of life, albeit one obscured beneath thick layers of cultural habit. Even on a ranch fractioned through property successions, the offspring of ranching families continued in the same occupation because it brought a riskless living. By killing their steers, such people had meat and fat at their disposal. By selling the hides of the same animals, they could buy salt, flour, *erva mate,* and even the simple clothing they needed. For the councillors of São Gabriel, ranching remained the preferred occupation "because there was nobody in the province who did not judge himself suitable for pastoral work"; and the estâncias therefore remained "infested" with agregados, individuals motivated by the prospect of food and by the certainty of a place to "set down" ("tendo certo o pouso").[19] In the language of modern development concepts, ranching was meeting basic needs.

These criticisms of the ranchers, made mainly by the politicians of the Campanha towns, had varied motivations. The positive aspect of their criticism came from a wish to replicate the success beginning to be associated with commercial agriculture further south. However, they gave little thought to what kind of physical or economic basis the ranchers worked with for planting crops. The educated had always viewed cultivation as a higher cultural state than ranching, and for some this was motive enough. But underlying the elite criticisms of the Campanha ranchers was the negative factor of a social fear about what would become of surplus labor. Displaced peons were a problem in the making. By 1883, the councillors at São Gabriel were already pessimistic that these people could turn to cultivation; they felt that cultural constraints, their "lack of vocation for work of this nature," and their widely held belief that crop farming was an inferior occupation made this unlikely. A reflection of this pessimism is the description of the greater part of the displaced peon population as living "as vagabonds under the vigilance of the police."[20]

Over the years immediately following, former ranch slaves joined the pool of displaced free peons in the Campanha.[21] At that point, local politicians were sufficiently concerned that they began to explore more seriously the prospect of linking surplus labor and cultivation. For example, in 1888 the councillors at Bagé suggested that local "material progress" would soon depend more on cultivation than on ranching, which they regarded as lying in a state of decadence. While in the middle of the century the council had shown no sustained interest in the prospect of colonization schemes for the region, its view had changed "with the complete extinction of the servile element." Concerned about the increasing number of people without clear occupation in the area—but offering no hint of how many—the councillors noted the "incontestable necessity and transcendent utility of developing agriculture." And they tried to loosen the provincial purse strings in order to fund a municipal colony.[22]

Experiments of this kind failed because they had nothing to do with the interests of the rural landowners at large. Throughout the period under study, the chief preoccupation of Campanha landowners was how to achieve the improvement of their ranches, a major task in a region with weak credit, poor transport, and uncertain future market growth. Despite the bold forecasts of ranching's imminent demise, coming by the 1880's not merely from the coast but also from the Campanha itself, it proved a remarkably durable form of land use. While the councillors of Bagé had begun to talk in 1888 of spending money on seeds and agricultural implements, that county did not actually undertake a large-scale experiment to diversify the character of land use before 1949.[23] Inherited social and cultural values, which still conferred status on ranching, broke down extremely slowly in a region which saw little challenge from newcomers.

Against this general picture of extremely slow change in cultural values, however, it must be remembered that from the 1880's onwards, some of the ranchers did change their habits, sometimes quite rapidly, as the economic rationale became clear. The fact that sheep followed Uruguayan railways to the Campanha frontier provides an excellent example. In addition, tariffs on imported food under the Old Republic soon led to large-scale experiments with rice cultivation. Those who battled the prevailing culture of conservatism to modernize ranching systems in the Campanha faced a series of structural problems not always shared by their southern neighbors.

Insecurity of Property

Insecurity of property comprised issues affecting relations both within and between social and economic classes. Progressive landowners were concerned about the legal guarantees attached to their property. They looked to central government authority for the protection of their land and livestock management rights, including against theft. While some of the issues required clarification of the rights of the individual rancher in relation to government and to neighboring landowners, ranchers' concerns about insecurity of property also included the matter of rural resistance to the concept of private property and the rights accompanying it. Displaced surplus labor, accustomed to a semi-nomadic existence that assumed at least vestiges of collective rights with regard to land use, fought back against the consolidation of individual property rights.

In the 1850's, there were still echoes of common property in widely held attitudes towards livestock in the Campanha (as discussed in Chapter 2). Over the remainder of the nineteenth century, the consolidation of private property was a vital issue for landowners on the southern South American grasslands. This did not occur at a uniform pace throughout the broad ranching region. Uruguay, on account of the weakness of the state, had been synonymous with insecurity of property for much of the nineteenth century. However, the implementation of rural laws during the 1870's gave the ranchers there a firmer basis for investing in property than their Campanha neighbors. Campanha ranchers contemplating innovations on their estâncias considered the legal guarantees backing major investments less satisfactory than those enjoyed by southern neighbors. The absence of a regional set of rules and regulations slowed the intensification of ranch production.

PROPERTY RIGHTS

As Warren Dean has shown, the Brazilian Empire had little control over land and was effective in its policies only in the coastal regions.[24] Imperial Brazil failed in its designs to prevent encroachment on public land, and failed therefore to prevent the extension of latifundia established without legal title. On the expanding frontiers of cultivation in south-central Brazil, such as the coffee plantations of São Paulo, this was a major issue.

In the Campanha, the context was different. Ranchers were less concerned with the extension of a system than with its intensification. Security of land tenure was weak throughout Brazil, but in the Campanha this was not the leading problem that it was elsewhere. Encroachment on public land did take place, helped especially by wire fencing (which could be very quick to materialize where the returns were great), but there were no vast new lands to occupy. The physical frontier of settlement in the Campanha had closed by the early nineteenth century; most of the region was granted as sesmarias before that system terminated in 1822. On the other hand, the weakness of state authority in the area of property rights was an important obstacle to the intensification of ranching. The key issue was not so much title to the land as the absence of general regulations supporting ranchers in the management of their property.

For some Campanha ranchers, answers to their problems were ready to hand further south. Uruguay first enforced its rural code in 1876, as the centralization of authority in Montevideo began to take hold. As Juan Oddone has argued, "Government slowly but inexorably secured a monopoly of physical force over the inhabitants of the country. For this achievement the large landowners composing the Rural Association . . . never ceased to give thanks."[25] In Rio Grande do Sul, ranching interests had only complaints about their governments, on account of the continuing insecurity of property, even into the Old Republic. While the national governments in Argentina and Uruguay passed regulations "creating and determining obligations for all of the rural population," as late as 1913 *A Estância* argued that the uncertain division of responsibilities between the federal and state governments in Brazil continued to slow this process.[26]

Campanha ranchers presented themselves as victims of politico-constitutional prerogatives, but this lack of clarity about the legal obligations for property management was a significant obstacle. For example, it slowed the development of sheep breeding. An article published in *A Estância* in 1913 related the history of the owner of an especially fine Campanha ranch, who, unable to continue its management, advertised his very large property for rent in Rioplatense newspapers. He attracted serious interest from a major company in Argentina and Uruguay, only to have the negotiations fall through because of the lack of any Brazilian law governing the ownership and management of sheep.[27] The absence of commonly understood legal norms (about the caliber of fencing required and the return of strays, for example) made the lease too risky. Many ranchers must have concluded that sheep were not worth the bother.

While the lack of comprehensive legislation on property rights undoubtedly slowed land-use innovations in the Campanha, it is difficult to establish to what degree it did so. Although piecemeal legal property guarantees backed by state authority were far from ideal for investment-minded ranchers, the adoption of wire fencing did much to increase the security of individual estâncias. Despite periodic complaints about the weakness of legal provisions for property management, those who understood the local milieu well and had some influence over its circumstances were probably prepared to invest without the backing of a rural code. On the other hand, the perceived deficiencies of legal guarantees regarding property can only have heightened the wariness of outside groups and helped to deflect the interest of important potential modernizers, including northwest European minorities, away from the Campanha. This carried major implications for the course of Campanha ranch modernization because these nuclei of "improvers" were responsible for disproportionate amounts of activity, a point identified both in the development theoretical literature and in earlier historical writing on the Plata.

OUTLAWS

In the colonial and early independent periods, state authority was widely dispersed over the vast geographical expanses of the southern South American grasslands. Governments were without the means to control these hinterlands and counted on powerful private citizens to do so. Commenting on Latin American cattle frontiers, Duncan Baretta and Markoff have argued that control by rural landowners over frontier populations provided the basis for much of the visible accumulation of wealth in the coastal cities. Economically successful ranchers needed to be both captains of war and diplomats.[28]

Again, Duncan Baretta and Markoff have argued that much of the population of the cattle frontiers was in some kind of flux, that "permanent nomadism and permanent geographical stability were two extremes, and that most of the population . . . belonged to neither."[29] Success in ranching depended on negotiated relationships with mobile labor. The wars around mid-century exacerbated the degree of rootlessness in Rio Grande do Sul, as was seen in Chapter 2. Although Campanha ranchers widely denounced vagrancy and complained about the lack of government protection from cattle thefts, many more probably tried to find a modus vivendi by allowing a few more agregados on their lands.[30] Striking a social contract with vagrants had its uses, especially at times of political instability. However, as the southern South American grasslands were drawn increasingly into the North Atlantic economy, the cultural character of Latin American cattle frontiers faced a dramatic transformation.[31]

Throughout the southern South American ranching region during this period, modernizing ranchers faced the problem of surplus labor on the estâncias. They had to control a swollen ranch population, one in the habit of seeing the rural round partly as an entertainment, as it evolved into a smaller, compliant core of wage labor. In both Argentina and Uruguay, a strengthened state apparatus supported property management rights. Some contemporary observers, including Brazilian ranchers resident in northern Uruguay, termed this process "authoritarian centralization." Their phrase was apt. Once dispersed through the hands of a series of mobile caudillos, authority had now become concentrated in the cities in the form of a codified law which the urban elites (many of them absentee landowners of rural properties) did their best to extend and make effective throughout the hinterlands. Gauchos who had once been encouraged to occupy corners of the estancias were now branded as bandits, malefactors, and vandals. Behind these epithets lay the great social costs of modernization.

The fate of marginal labor differed according to the uneven geography of repression of the gauchos. And the impact of labor legislation varied according to the socioeconomic conditions prevailing at the time it was implemented. For example, in Buenos Aires Province, where in 1880 there was still an active ranching frontier of settlement, the resolution of social tensions focused on the margins. Through the labor repression that marked the rural code, landowners saw local vagrants (displaced peons) drafted into the army to fight Indians.

By contrast, Uruguay no longer had new land for ranchers to take up. This country was essentially a closed system, with the notable exception of its porous northern frontier with southern Brazil. In the wording of the day, the dictatorial government in Uruguay worked to make its territory "habitable." It did not shrink from using social repression to achieve that aim, as contemporary observers were aware. Eduardo Acevedo, the liberal historian, noted that the newspapers of Uruguay's interior departments did not let a week pass without recording the death of people held in police custody.[32] And in an edition of their great compendium *Handbook of the River Plate*, the prolific Mulhalls noted rather tersely that a Scot, Major McEachen, *intendente* (governor) of Paysandú from 1873 to 1877, had "cleared the department of banditti."[33] For some Brazilian ranchers with land in Uruguay, these developments were welcome. General Osório, a national hero and part of the Riograndense elite that held most of its assets south of the border, argued before the Brazilian senate that "malefactors" would not be remaining long in Uruguay. The government there was "pursuing the vandals," with the result that they were "now taking refuge on Brazilian land."[34]

Social conflict in Uruguay carried implications for Rio Grande do Sul. As the wave of fencing diffused through Uruguay, many of the displaced peons sought refuge in the hillier, northern sections of the country along the Campanha frontier. These peons form a classic example of what Burns has termed "victims of the state."[35] Without other visible means of support, many turned to banditry.[36] Thus if Uruguay's Paysandú department had been cleared in the 1870's of those without land or work, the Mulhalls noted in the same 1892 edition of their guide that the woods of the huge department of Cerro Largo along the Brazilian frontier were "much infested with banditti."[37] Even the ecology of the international frontier zone fostered resistance.

The cultural conflict that accompanied fencing and the rural code in Uruguay stirred up concern about property management rights in the Campanha. Ranchers there were fully apprised of what was taking place further south; some of them wanted legal powers similar to those granted to their neighbors in Uruguay, particularly a sanction to throw marginal labor off their land. In 1877, the Alegrete councillors described the absence of guarantees of property, in the face of constant cattle thefts, as one of the two main impediments to development in their county (the other was degeneration of the creole breeds of livestock) and called for the organization of a rural police force, "under whose severe vigilance" the property rights of the local inhabitants would repose.[38] Similar concerns about outlawry were raised in Jaguarão, where the councillors asked the provincial president twice for permission to form a private rural police force.

During 1877, the provincial assembly took up concerns about lack of rural policing as an obstacle to development on behalf of the Jaguarão councillors, but they became enmeshed in party politics. Deputy José Francisco Diana, a Liberal politician, complained that the police chief in the town of Jaguarão (expressively described as "inertia personified") left "malefactors" at their ease on the grasslands. Diana denounced the poor quality of provincial administration, complaining that Rio Grande do Sul had been "flagellated by the satraps" sent down from Rio de Janeiro, who were "indifferent, if not hostile" to the regional well-being.

He brought his speech to its culmination with the striking observation that a Jaguarão rancher had preferred to rent land in Uruguay, even during the civil war there of the early 1870's, rather than see the daily erosion of his animal wealth through thefts within Rio Grande do Sul.[39]

The call from Jaguarão for rural and not merely urban policing came at a time when Liberals controlled the provincial assembly but the president was a Conservative. While the Liberals made political capital against the Empire with this case, they achieved little by way of changing the basis of rural law in the Campanha. Even as the Empire drew to its close, there were complaints about the limitations of rural law in relation to outlaws. Public security had become very different in the towns of the Campanha than in the strictly rural areas. If town life had been calm during 1888, the Bagé council noted, "security of life and property" remained uncertain in the rural parts of the county, mainly on account of bandits, "enemies of work and domestic tranquillity," who sometimes "infested" the frontier zones.[40]

Despite eloquent pleas, the Empire was simply not interested in the devolution of legal powers to ease the adoption of technology by Riograndense ranchers, a marginal political group. Thus, the Campanha forms a case where (borrowing from Burns's terminology) the "folk" had some power to mediate change, the modernization of the estância, over a longer span of time.[41] It is also clear that the absence of draconian labor legislation of the type employed in the Río de la Plata made the social turmoil associated with fencing Riograndense estâncias a long process of attrition. The absorption of marginal ranch labor in the Campanha is largely undocumented, but speculation suggests some likely escape routes from destitution. During both the civil war in Rio Grande do Sul of 1893–95 and Saravia's revolt from northern Uruguay of 1903–4, all sides contracted these people as mercenary soldiers. Many fell in the battles of the 1890's, when there were ten to twelve thousand casualties, equivalent to roughly 5 percent of the Campanha's population.[42] Others probably found their way into the labor forces of the newly appearing slaughtering plants of the interior towns; in 1918 the Armour frigorífico of Livramento alone employed 1,380 workers, 50 percent Brazilian and 40 percent Uruguayan.[43] Still others drifted out of the Campanha, either toward the towns of the Litoral or even beyond the state. Even so, in the Campanha's international frontier zones, banditry remained the first option for some. Modernizing ranchers, especially the sheep breeders of Quaraí and Livramento along the Uruguayan frontier, were still feeling its effects as late as 1915.[44]

The legal basis of rural property management in Rio Grande do Sul was very different from that in the Plata. As Duncan Baretta and Mark-off have pointed out, "Rio Grande do Sul . . . long deprived of political power within a large and economically differentiated nation . . . did not exhibit the variety of governmental measures to control ranchhands that were used in Argentina."[45] More important still, the Campanha ranchers did not share the legal backing enjoyed by their immediate neighbors in Uruguay. Without these measures, it took several decades to absorb the surplus rural labor produced by fencing. The "folk" resisted. And modernization of the ranches was slowed as a result.

Infrastructure Problems

THE PORT OF RIO GRANDE

Throughout most of the period under study, Rio Grande do Sul confronted a major infrastructure problem in the poor quality of its port. Already a source of concern at mid-century, this difficulty only worsened in the following decades, as the character of shipping underwent drastic change with the advent of steamships and larger sailing vessels.[46]

The other ports that served the pastoral regions of South America were inefficient to a different degree.[47] The natural harbor of Montevideo was much deeper than Rio Grande's. Before the Argentineans made massive investments in new port works at Buenos Aires, Montevideo was by far the best natural harbor of the broader region. British consular reports provide a clear sense of the increasing marginality of the port of Rio Grande relative to the Río de la Plata for British shipping.[48]

The problems of the port at Rio Grande emerged very clearly during local political debates of the early 1860's, when the traditional markets for ranch products were oversupplied.[49] Most of the foreign market for southern Brazil's port lay with Britain and the United States. To cross the sandbar at the entrance to the port, the merchants of these nations were forced to charter ships of limited draft, which in turn increased shipping costs. Local politicians claimed that insurance costs were also higher than around the Plata. British insurance companies assumed responsibility for consignments of hides as soon as they were loaded in Pelotas. Such consignments faced the costs and risks of being moved by light vessel from Pelotas through internal waters to the ports of Rio Grande or São José do Norte for loading onto oceangoing vessels, which then had to negotiate the bar to reach the open sea. It was not only international trade that was affected by these navigation obstacles.

Larger Brazilian vessels involved in the coastal trade began to shun Rio Grande in favor of Montevideo.

The prosperity of Rio Grande do Sul depended upon the relationship between pastoral exports and foreign trade. At Rio Grande, foreign merchants drew from a more restricted hinterland than at Montevideo. This made it more difficult to achieve the same degree of balance between imports and exports, a factor which further drove up shipping costs. When merchants could not find what they sought at Rio Grande (mainly hides), they transferred their attentions to the Plata.

This gap between Rio Grande and the Rioplatense ports, already important by the 1860's, only widened in subsequent decades. Expensive interruptions to trade on account of the condition of the bar became more frequent.[50] Despite organized dredging and pilot services, the silting of the channel at the entrance to the Atlantic had intensified. No single indicator of the pressing need for harbor works in the south during this period could be more eloquent than the following: in 1880, freight coming from Europe paid around 80 shillings per ton at Rio Grande, as opposed to 25 at Montevideo.[51] With this huge difference, it was small wonder that Campanha ranchers found Uruguayan contraband goods appealing and that British shipping lines decided that "the [Rio Grande] trade no longer suited them."[52]

In the 1880's, the commercial associations of Rio Grande do Sul lobbied in concert about the need for extensive port improvement at Rio Grande, one of the very few issues they were able to agree on.[53] However compelling the case for a solution to the problem of the bar, the issue did not move far beyond the drafting table during the Empire. There was nothing anomalous about this within the national context of Brazil. Slowness in the provision of infrastructure was one of the Empire's defining characteristics, and Rio Grande do Sul did not stand alone in needing port improvements in order to stimulate export development. Money was spent on elaborating solutions. There is a considerable literature of technical reports written by a range of engineers, Brazilian and European (including British, French, and Dutch). Such documents languished. Part of the reason lay in a lack of clarity about means.[54]

The lag between identifying the character of port problems and commencing major public works was usually long throughout the broader region. While the technicalities of improving the port at Buenos Aires were extensively debated from the 1860's onwards, the first phase of the modern docks there was only completed during the 1890's.[55] Work to provide Uruguay with a modern port at Montevideo did not begin until

1901.[56] Nevertheless, the South American agenda for infrastructure improvement was set in the Plata.

The Brazilian government could not remain blind to these vast turn-of-the-century developments. As the country looked to be more prosperous, with a rising trade surplus and an influx of foreign investment, the government began to invest in long-neglected public works.[57] For example, it authorized major modifications to the port of Rio de Janeiro after 1904. In 1906, a concession to clear the bar at Rio Grande and to develop a modern port was granted to an American, Elmer L. Corthell, the same engineer who had advised on modifications to the port of Buenos Aires in 1902. Corthell had difficulty raising the necessary capital, and in 1908 the concession was transferred to the Compagnie Française du Port du Rio Grande do Sul. This in turn became one of the 38 subsidiary parts of Percival Farquhar's syndicate, the Brazil Railway Company.[58]

Corthell's technical solution for the Rio Grande bar, the construction of two vast breakwaters, differed only in detail from the broad ideas presented decades earlier. Fear of cost had been a more effective obstacle than technology. The scale of the Rio Grande project was such that it was only completed during World War I. However, the works in progress were beginning to make their impact even before the war. Traffic at Rio Grande more than doubled between 1902 and 1912 (from 235,817 to 523,359 tons); in the latter year, British shipping predominated for the first time in years.[59]

In the first decade of the twentieth century, rates for ocean freight at Rio Grande benefited from the massive expansion in world shipping supply. Yet elements of the port's marginal status remained. The quantities of shipping handled by Rio Grande and the Río de la Plata ports were still on totally different scales; Buenos Aires handled 18,000,000 tons in 1910. By the outbreak of the war, the large numbers of emigrants streaming onto the pampas and the successful development of a refrigerated-meat business had transformed the Río de la Plata into one of the major markets of the world for liner services. At Rio Grande, sail was not yet completely obsolete. Schooners (mainly Scandinavian) still dominated the salt trade from Cádiz, exchanging their cargo at Rio Grande for hides.[60]

By World War I, the facilities at Rio Grande had seen massive improvement, and the technical weaknesses were largely addressed. As was the case further south, a modern port paved the way for the establishment of frigoríficos. Yet it was not only a good port that was needed, but one with reasonable operating costs. In this respect Rio Grande remained a problem for the ranching economy.[61] High port charges reflected the expense of the massive engineering works to clear the bar. Most of the American frigoríficos established in Rio Grande do Sul avoided it. Even the domestically owned frigorífico relocated its projected site from Rio Grande to Pelotas to avoid the high port handling charges. Throughout the 1920's, Montevideo remained a highly effective competitor for shipping Campanha production.[62]

RAILWAYS

The poor quality of the port during the second half of the nineteenth century reduced the potential contribution of railways in the Campanha. The relatively late arrival of the railways reflected lack of conviction on the part of decision-makers in south-central Brazil about their economic utility in areas of sparse population. Even local enthusiasm along the southern frontier was mixed when the economic returns seemed questionable. While provincial president Marcelino de Souza Gonzaga noted in his report for 1865 that he could see the strategic and political values of linking "the frontiers to the more civilized and commercial centers of the province," he raised a series of doubts about whether the railway would have any serious impact either on the pastoral industry or on the locations of the charqueadas.[63] Two years earlier, spatially conscious British Acting Consul Alexander Gollan had offered his opinion that "opening up the country by means of ordinary roads intersecting each other would be more generally beneficial than a railway, the advantages of which would to a great extent be confined to one district."[64] Politicians from other areas of Brazil would have agreed.

After the Paraguayan War, the strategic motive for building railways across southern Brazil needed no defenders, but belief in the economic advantages of doing so was still limited, especially on the part of the imperial government. Funds set aside during 1872 for railway construction in Rio Grande do Sul instead went to cover more immediate needs. Even as the project gained firmer official support towards the end of the decade, Consul Gollan, back for another tour of duty in Rio Grande, did what he could to dissuade British investors from sinking money in the project.[65]

The economic basis for railway construction was not easily identified in the Plata, either, in its early stages. The early lines received an important boost from merchants and landowners, including the British.[66] With little push from the mercantile elites in Rio Grande do Sul, little happened until nearly the end of the century. When Brazil did develop a skeletal network for the Campanha, its motivation was as much to counter the Uruguayan network already in existence as to plan routes and related infrastructure to enhance the mobility of the inhabitants of the province. In the 1890's, as the lightly engineered trunk lines were being readied for traffic in the Campanha, Uruguay floated plans that sought to extend that country's standard-gauge system far beyond the Brazilian frontier into the heart of South America.[67]

Uruguay held a clear advantage in railway development; it was easier to think in terms of a national transport plan in a small country. However weak their potential initial traffic returns in the Campanha, the imperial government could not afford to ignore its neighbor's efforts to integrate space with railways. That it did ignore them demonstrates how the Campanha suffered from its peripheral location within Brazil.

FINANCE

In the context of the broader pastoral region, banking was but a third aspect of weak infrastructure in the Campanha. The evidence is fragmentary, but it is most unlikely that ranchers gained access to any formal long-term credit under the Empire. An extremely poor circulation of rural credit was one of the most distinctive features of imperial Brazil. The foreign banks were more interested in garnering profits from smoothing the import-export trade than in providing long-term credit to landowners; the rationale behind the early branch of the London and Brazilian Bank in Rio Grande do Sul, opened at Rio Grande in 1863, was that it formed the "natural complement" to the branch in Recife. Branches located in zones with different economic geographies helped to balance drawings and remittances.[68]

In contrast, the efforts of the Barão de Mauá, the great figure in mid-nineteenth-century Brazilian finance, included elements of development banking. He paid special attention in his banking ambitions to his home province of Rio Grande do Sul, where the branches founded in the Litoral were part of his network for the Río de la Plata. In 1860, Mauá informed Ricardo José Ribeiro, his partner and manager in Montevideo, that he wanted his banking philosophy advertised in Rio Grande do Sul "by means of circulars and other published notices, so as, in time, to establish a branch or agency of Mauá and Company in every town or city in the province with a population of three thousand or more."[69] Ribeiro had already been active in this regard, drawing specific attention in an advertisement from Rio Grande to the value of letters of credit for cattle buyers exposed to the risks of traveling in the interior of Uruguay.[70] Despite Mauá's attempts to broaden the social base of bank users in Rio Grande do Sul, inventories suggest that only the wealthiest ranchers made use of his branches, and even then only to a limited degree. In any case, it was not until the 1880's that ranchers had a general need for investment in their properties, and by then Mauá was already a spent force as a banker, a victim of the financial strains of the Paraguayan War on the Empire. The modest São Gabriel savings institution started in 1853 (see Chapter 2) met a similar fate.

The gap between southern Brazil and the Plata in the supply of rural credit widened after the Paraguayan War. Both Argentina and Uruguay formed their first mortgage banks during the 1870's. After 1881, British mortgage companies moved into Argentina and Uruguay. But they were very selective in their credit geography with ranchers, restricting "their lending to the very best pastoral areas of the provinces of Buenos Aires, Santa Fé and Entre Rios, and of Uruguay."[71] Such developments in finance in the Plata circumvented the Campanha, where, during the late Empire, ranchers remained starved of formal credit.

A rapid expansion of the banking network throughout Brazil accompanied the establishment of the Old Republic. The number of banks operating in Rio Grande do Sul increased from two to ten in 1890–91 alone. By the 1890's, bank branches were common in Campanha towns. By World War I there was no lack of financial institutions in Rio Grande do Sul. The spatial pattern of these institutions is of less interest than their character; it remains unclear whether banking with a development and not merely a commercial orientation was taking place. The federal government laid plans for a mortgage bank similar to those in Argentina and Uruguay, but this never moved beyond the statute book. Mortgages were much less common in Brazil than in Argentina.[72]

In his research on British financial institutions in Argentina, Charles Jones has observed that these became more amenable in their treatment of the landowners in the decade before World War I.[73] Money also began to flow more freely in Brazil in the same period. In his major study of Brazil's political economy, Topik has charted the limited state endeavors to supply rural credit in Minas Gerais and São Paulo but offers no comment on Rio Grande do Sul, the third leading state of the nation.[74]

In Rio Grande do Sul the state government took less initiative to help landowners than did its counterparts in south-central Brazil. Ranchers did make direct demands of the state for credit, but they were blocked by the character of the government. After the civil war of 1893–95, the state government was rigidly controlled by the authoritarian political machine of the Partido Republicano Riograndense for the remainder of the Old Republic. Imbued with the positivist ideology of the PRR, the state government declared that it was not prepared to help any particular sector of the economy directly.[75] The lack of state-backed credit generally became a difficult structural obstacle for Campanha ranchers. This pattern did relax temporarily with the economic expansion of the state during World War I. With the economic euphoria of the war, ranchers gained access to credit, especially through the Banco Pelotense, a state-supported institution that expanded its activities greatly.[76] After the war, the political doctrines of the PRR stood in the way again, when ranching credit became confused with the railway issue.

The Farquhar syndicate had barely begun to make vital improvements to the southern Brazilian railway system before the First World War cut its supply of European finance. Ill-equipped from the outset, the regional railway system performed miserably during the economic expansion and inflation of the war. A major strike of railway workers paralyzed the state for part of 1917.[77] Influenced by these matters and by its conviction that transport belonged under public management, the state government worked to achieve regional control over the port of Rio Grande and the railways. For the PRR, this had a greater priority than obstacles to rural credit, and it met its goals. The federal government transferred authority over the port to the state in 1919 and authority over the railway system in 1920.

The Campanha paid much of the price for this. To secure funds for transport improvement in 1920–21, the state government withdrew money from the Banco Pelotense, which in turn placed that institution (and therefore ranching credit) in crisis.[78] Rural credit was a major issue during the recession of the early 1920's that followed. Assis Brasil, in his address to the Ranching Congress held at Porto Alegre in 1922, called for the large-scale extension of federal credit to the pastoral industry and for the multiplication of Bank of Brazil branches in the ranching zones.[79] However, his concern for the diffusion of credit came to nothing at the time; Assis Brasil was in opposition rather than in government, and Rio Grande do Sul was drifting toward civil war. The weakness of ranching credit was of more than passing significance in the causes of the revolution of 1923.[80]

In Argentina, a current research focus is the ranchers' monopolization of credit at the expense of farmers; in contrast, the provision of formal credit to Campanha ranchers was generally poor before the later 1920's.[81] The absence of what a British official had termed in 1919 a "proper organization of land banking" was widely viewed as one of Rio Grande do Sul's leading development obstacles.[82] Even in 1920, this particularly weak element of the infrastructure was still hindering the modernization of ranching. The consequences of this weakness became more fully apparent with the erratic performance of the region's frozen-beef exports during the early 1920's.[83] Significant change came only with the pragmatic administration of Getúlio Vargas. Mindful of the importance of local ranchers' support in his bid for national power, in 1928 he used his gubernatorial authority to set up the Bank of Rio Grande do Sul, an institution designed specifically to funnel cheap credit to rural producers.[84]

Cities, Merchants, and State Policies

METROPOLITANISM

Metropolitanism played a central role in the drama of rural transformation in southern South America. By the opening of the twentieth century, Buenos Aires and Montevideo had achieved a striking degree of urban primacy within their respective countries. As large commercial-bureaucratic centers, they served as the relay points for technological innovation on the grasslands. Almost all of the elements in the chemistry of modernization were to be found within these cities, both of which were backed by immediate hinterlands of great agricultural potential. Argentinean and Uruguayan ranchers who had tired of life in the camp had no difficulty recognizing Buenos Aires and Montevideo as the centers of national life. Finance, politics, and social life coalesced in these major port-cities of the estuary. These were already centers of great regional importance in 1850, and the development of the Río de la Plata for the remainder of our period can be written in large part around the theme of metropolitanism. Railways, telegraphs, and professional armies equipped with something more effective than lances reinforced the hold of Buenos Aires and Montevideo over their national territories and even beyond. How different all of this was from the Campanha, where there was no single city with the power to meet the diverse needs of ranchers and other inhabitants.

The ranchers of southernmost Brazil were torn between competing metropolitan centers. Ranching business, reinforced by the pattern of the Uruguayan railway system, drew Riograndense estancieiros toward

Fig. 7.1. The social face of Europeanization: *footing* on the pier at Pocitos Beach, Montevideo, c. 1900. Photograph by Chute & Brooks. Reproduced from Hoffenberg, *Nineteenth-Century South America in Photographs*, p. 109.

the Rioplatense cities. But central political power resided in Rio de Janeiro, and most of the ranching elite who received professional education went north for it, especially to the São Paulo and Recife law schools. Until World War I, when the railway link with São Paulo was completed, even south-central Brazil lay a sea-voyage away. For those who stayed within Rio Grande do Sul itself, the pull of the city was fragmented among three centers, all small and all at some remove from the Campanha grasslands. As late as 1900, the population of Porto Alegre was still only 74,000, although growing rapidly.[85] By comparison, Buenos Aires was a true metropolis, home to over 1,300,000 people by 1910.[86] The situation also differed from that of the Plata in that metropolitan influence within Rio Grande do Sul shifted over time. From roughly 1820 to 1860 the center of regional economic strength was passing southward to Pelotas and Rio Grande with the rise of the "traditional" ranching economy. By the opening decades of the Old Republic, the focus was becoming much more concentrated around Porto Alegre, based largely on the commercial transformation of the colonial zones.[87]

Despite their proximity, Rio Grande and Pelotas had very dissimilar social and economic profiles. The immediate hinterland of Rio Grande, surrounded by sand, held limited economic promise. Yet for all its natural weaknesses, the town's location near the junction of the inland waters and the waves of the Atlantic made for trade. Foreign merchants predominated in the commercial structure of the port.[88] Some of the most prominent concerns were branches of Río de la Plata mercantile houses. The British mercantile community had strengthened at Rio Grande around mid-century with the instabilities of the Rosas era further south.

While the interests of some of the domestic elites were divided between the two towns, prosperous Pelotas was the favored place of residence of what even the Conde d'Eu found himself able to call "the Riograndense aristocracy."[89] During the Empire, domestic elites controlled almost all of the wealth generated by the charqueadas. Pelotas attracted very few foreign merchants. British consular registration data for 1891 show only 3 percent of local Britons living in the town.[90]

By 1880, the commercial elite of Pelotas had made significant investments in improved infrastructure, but mainly at a very local scale. These included such ventures as the canalization of the Santa Bárbara River, as well as the provision of paving, gas, and piped water. After years of effort the leading industrialists had finally opened the São Gonçalo River to international shipping in 1876, to free themselves from a dependence on transferring ranch products from launch to ship at the ports of São José and Rio Grande.[91]

Once foreign ships could reach Pelotas directly from the open sea, a struggle with the merchant community of Rio Grande ensued. The provincial president had increased the powers of the Pelotas tax-office in 1880. The merchants of Rio Grande wanted these measures rescinded, while their counterparts at Pelotas were looking for still more local fiscal power. In arguing their case before the Empire, the merchants of Pelotas adopted a tone of clear moral superiority. They complained that Rio Grande was a protected settlement that had stood out against "the laws of free commerce and industry."[92] While in Pelotas commercial profits were visibly invested in public works, profits earned in the import-export trade at Rio Grande had a "mysterious destiny." Foreign merchants passed quickly through the entrepôt (*entreposto*) of Rio Grande and took their profits home.[93] A town like Pelotas, "opulent and entrepreneurial," had no need for "the annihilation of a sister town in order to progress."[94] Despite that proud assertion, Pelotas merchants signed an agreement at the beginning of 1880, refusing to buy goods purchased for resale in the Campanha brought into the province through the hands of the foreigners in Rio Grande. This aroused great suspicion on the part of the ranchers of the Campanha, who viewed this protective measure by the Pelotas elite as an effort to "enslave our pastoral industry, reduce our breeders to the character of simple delivery-men to the charqueadas."[95]

In hindsight, it is easy to see that rivalry between separate groups of merchants in Pelotas and Rio Grande represented a dissipation of development effort. Although achieved with considerable local effort and expense, the fact that small ships could come from the Atlantic directly to the sites of the Pelotas charqueadas by the 1880's was of limited significance. When the earliest experiments with refrigerated-meat shipments to Europe were already taking place further south, the key infrastructure constraint was not the character of internal navigation but the increasing marginality of the bar.

MERCHANTS

In the areas of most visible material progress in southern South American ranching before World War I, the path of economic development involved a very tight correspondence of interests between foreign merchant capital and the domestic elites. Around Buenos Aires, for example, this produced the distinctive commercial-bureaucratic city that James Scobie has described. The pattern was different for the Campanha. There were linkages between foreign merchants and the modernization of ranching, but they were vastly weaker in Rio Grande do

Sul than in the Plata, a pattern bearing out Friedmann's theory that interaction tends to be more intense within large city systems than in peripheral settings.[96]

When Buenos Aires was still termed the *gran aldea* (large village) and Montevideo was barely that, they were of more potential interest to foreign merchants than Rio Grande on account of their locations alone. After all, in the Plata estuary they stood at the gateway to the major river systems leading to the heart of the subcontinent. Beyond these ever-important pragmatic considerations, the style of life also had more to offer. Even at mid-century, Buenos Aires or Montevideo would have suited most Europeans better than the sandbanks and isolation of smaller Rio Grande; for example, the French scientist Bonpland, after a visit to a leading mercantile house at Rio Grande in 1849, observed that while the company had been agreeable, social intercourse was usually limited; "what a difference," he noted, "from the dinners of Montevideo, the customs, the service."[97] Even during one of its most flourishing phases, Rio Grande was barely on the social map.

Although the remnants of the British mercantile community at Rio Grande continued to be of local economic importance until the early 1930's, their decline began long before.[98] The increasing difficulties of the port in the second half of the nineteenth century reduced the appeal of Rio Grande as a place to conduct business. As the pace of development quickened around Porto Alegre, where the commercial agriculture of the colonial zones was already leading to industry, the British mercantile community began to move northward. Competition with the contraband trade had always been a problem for those who paid their import taxes at Rio Grande, but it intensified with the extension of the Uruguayan railway system towards the Campanha frontier. An honest merchant was better placed in the Río de la Plata, where the British mercantile communities also strengthened throughout our period; their interest in servicing the needs of the pastoral economy grew very rapidly in the opening decades of the twentieth century.

British merchants had begun to invest in estancias in Argentina and Uruguay quite early in the nineteenth century.[99] They bought land and sought a return on their capital, though not necessarily an immediate one. Direct mercantile involvement had an important impact on the adoption of innovations. In the early stages of many of the ranching innovations, foreign-born minorities were never far away. Northwest European minorities had a high profile in the rural modernization in the Río de la Plata. Such British surnames as Drabble, Gibson, Hughes, and

Jackson are still immediately recognized in the Plata for their connection with ranch modernization. By no means all of the best-known foreign entrepreneurs were British. For example, Domingo Ordoñana, one of the founders of the Uruguayan Rural Association, and its permanent secretary after 1875, was a Basque immigrant.[100] Foreigners of diverse backgrounds, many of them monied from the outset, formed the cores of modernizing groups. Establishing islands of nineteenth-century "progress" on their lands, they slowly rattled the tenets of an ultraconservative rural society through the demonstration of alternative "genres de vie."

Small and isolated, the British mercantile community at Rio Grande remained almost exclusively urban. It did not enter the fabric of the ranching elite in anything approaching the same fashion as around the Plata estuary. In the years before World War I, there were still around six hundred British subjects in the interior of Uruguay, compared with a handful in the Campanha.[101] This is a striking difference, considering the broad similarities in land use of the two economies. The British role in the modernization of southern Brazilian ranching was slim. Compare an isolated reference to a company importing livestock into Rio Grande do Sul to the situation in the Plata, where one recent study has found that "all the major British mercantile houses appear to have been willing to import livestock."[102] No British mercantile firm at Rio Grande compared with Drabble Brothers of Buenos Aires and Montevideo, which engaged in cross-cultural transfer of technology as early as the 1870's by explaining the economic advantages of Australian fencing methods to Uruguayan ranchers in the journal of their Rural Association.[103]

The stimulus of outsiders was largely missing in the Campanha. There, the modernizers emerged later and from local ranks, linked with education. As education became more popular for the landowning elite, the cities took from the ranches their most illustrious sons.[104] Nobody could illustrate this better than Joaquim Francisco de Assis Brasil (1857–1938), whose life spanned a childhood on an estância in São Gabriel and statesmanly involvement in world economic conferences during the early 1930's.

Born into an affluent Campanha ranching family, Assis Brasil received his education in Pelotas and at the São Paulo Law School. While he inherited estâncias, Assis Brasil was never solely a rancher. Concerns for rural development and politics intermingled throughout a long and distinguished public career. Disenchanted by the character of local politics at the outset of the Old Republic, Assis Brasil entered the diplomatic service, which provided both an opportunity to witness ranching developments in other countries and probably some of the necessary time for reflection about what he had seen. For example, when serving in Washington as minister to the United States at the century's end, one of the first serious questions Assis Brasil asked of his country was why it was not monitoring the agricultural progress of the United States more closely. If Germany and Russia could post agricultural attachés to the largest single producer in the world, why not Brazil, whose need for observation was so much greater?[105]

Although he did not end his diplomatic career until 1912, Assis Brasil devoted as much time as possible to practical rural development in the Campanha well before then. In 1904, he bought land for his residence and model establishment, Pedras Altas. It became something of a mecca for rural progressives throughout Brazil and was highly regarded in the Plata. Assis Brasil wrote extensively about his practical ranching experiences, going so far as to publish a description of Pedras Altas, illustrated in part with plates of famous horses bred in the stud of the Dukes of Westminster.[106]

By contrast, many of the most prominent rural entrepreneurs in the Río de la Plata (often British) had begun as merchants. While aware of the status conferred by their land, these people were driven mainly by material concerns. Against the practical initiatives of these cores of "improving" enthusiasts, the Campanha could point to the work of a sole rich Brazilian of Anglophile tendencies. And while Assis Brasil was serious about material questions, he was also capable of laying them aside in his passion for politics. Earl Richard Downes has argued that Assis Brasil was the leading regional politician in Brazil advocating agricultural diversification through intensification.[107] What Assis Brasil managed to do with his own properties is extremely interesting, but nobody would argue that his message transformed the national agricultural scene. Assis Brasil only gained formal national influence as agriculture minister in Vargas's provisional government of the early 1930's, by which time the federal coffers were empty.

POLITICAL ECONOMY

In the middle decades of the nineteenth century, Campanha ranchers were often told by the Empire's supporters that the political structures of Brazil served them well.[108] Commentators given to litanies of the inherent weaknesses of Campanha ranching in comparison with that of the Plata (the poorer quality of the grasslands and of the port, for example) usually fell back on the strength of Brazil's government institutions as a last line of defense of the Empire. They warned Riograndenses not to exchange the "safety and order" of Brazil for the greater productivity of the old Cisplatine.[109] Campanha ranchers were unlikely to see their wealth decimated from one day to the next by the arbitrary acts of caudillismo, as in Uruguay, so the argument ran.

After about 1870, however, with the consolidation of Argentina and Uruguay as nation-states, the arguments in support of the Empire rang increasingly hollow. In the Río de la Plata, ranching was the dominant element in the economy; lobbying the government could—and did— bring quick rewards. This was the rationale behind the much earlier foundation of the agricultural societies there. Rioplatense ranchers, anxious to develop their properties, molded national politics to reflect their sectoral interests. From the period of its formation in 1866 to 1914, members of the Argentine Rural Society supplied over half of the country's presidents and 39 of 93 cabinet appointments.[110]

The ranchers of Rio Grande do Sul had little chance of following this pattern. In fact, until the end of the nineteenth century, politics stood in the way of rural organization in Rio Grande do Sul, which was but one peripheral portion of a vast country with an economy producing relatively little that entered international markets. Ranchers could have expected little from lobbying in Rio de Janeiro. In fact, many of them had a deep distrust of central power. Rio de Janeiro could and did appeal to nationalism on the southern frontier at times of emergency, but the reach of central government was little welcomed in the coxilhas of the Campanha. Earlier-nineteenth-century issues, such as the loss of the Cisplatine and the Farroupilha, lingered in the regional psyche as a continuing distrust of outside power. The Campanha was a stronghold of Liberal Party politics and home to some of the leading critics of the Empire.

Many instances of the consequences of this political marginality for the modernization of Rio Grande do Sul's ranching have already been seen, as, for example, in the account of the slow diffusion of barbed-wire fencing and railways. Political marginality also affected the region's development in other important areas, such as tariffs and taxation.

Repeated calls for a special tariff regime along the southern frontier only bore fruit after 1878, when the regional politician Silveira Martins was in charge of Brazil's national treasury. Even then, Uruguay, in its concern for national development, easily undercut much of the force of lower tariffs. The newspapers of the late Empire are rife with reports of the severity of the extent of contraband; one of the inspectors of the customhouse at Uruguaiana informed the minister of finance that the smugglers of goods from Uruguay were so organized in his town that they might be more accurately termed a "company" than an "association."[111] Despite the Empire's belated recognition that the Campanha needed help with tariffs (the extended special tariff of 1888 came close to creating a kind of de facto economic federalism), the advent of the Old Republic was marked by the decision to bring Rio Grande do Sul back into line with the remainder of the country.[112]

Taxation served Campanha ranching poorly in comparison with that in the Plata. Sensitive to the power of the landowners, imperial Brazil raised most of its inadequate income through the customhouses. Although a concept unpopular with landowners, direct taxation of land and/or its production could have been designed to encourage innovation on the estâncias. The idea of a land tax was being aired by 1860 in Rio Grande do Sul, but the provincial deputies decided in favor of adjusting the taxes already in place. Aware that new taxes on land were bound to be expensive to collect, as well as a source of irritation to the influential landed interests, they refrained from making major reforms.[113]

Uruguay, on the other hand, in dire financial straits after the Guerra Grande, had levied a direct tax on land and cattle as early as 1853. Granted, the Uruguayan state had neither the political power nor the bureaucratic means to collect much from it at that time, but the decision was still very important.[114] The unpopular idea of taxing land had at least been broached. As Uruguay became more centralized during the 1870's, this land tax became a serious source of revenue. The government also used it as an instrument to foster ranch modernization. By this time, Brazilian officials were aware of the potential value of direct taxation in stimulating the development of rural properties and were laying the theoretical groundwork for such taxation. Thus, in 1878, the provincial president of Rio Grande do Sul supported the establishment of a property registry maintained at the county level as a useful source of information in the event that a land tax came into being.[115] Karl von Koseritz (the first politician drawn from the ranks of the nineteenth-century colonists in Rio Grande do Sul) made a formal proposal for a modest land tax in 1883, his first year in the provincial assembly.[116] But during the Empire nobody was prepared to confront landowners over such schemes, in Rio Grande do Sul or elsewhere in Brazil's vastness.

Transition from monarchy to republic was a bloody affair in Rio

Grande do Sul. The subsequent authoritarian state government of the PRR displayed fewer qualms about political opposition from the Campanha than had the Empire. The ranching elite of the late Empire, with property concentrated in the Campanha and in northern Uruguay, was largely on the wrong side of the PRR. In 1904, the state government introduced a rural property tax which supplanted the provincial export taxes. With the inflation of land prices during World War I, it became the leading source of revenue for the state.[117]

Taxing land brought increased expectations of government. Perceived high taxes on land were hard to accept in the absence of a rural police, a rural code or a network of roads and bridges to ease the transit of goods. The land tax was inevitably interpreted within the Campanha as favoring urban property owners.[118] Ranchers subjected it to detailed criticism, complaining especially that the levy was on their capital (the estimated sale value of their land) and not on the more volatile value of their ranch production.

The whole subject of the design and impact of Rio Grande do Sul's land tax still awaits detailed research. But the very fact of the tax measure of 1904 must have led to increased concerns about profits and likely pushed the ranchers forward in their adoption of technology.

The most astute regional politicians had no difficulty recognizing that Brazil's systems of monoculture and economic dependency hampered modernization. While Brazil had all the political apparatus of a nation, in economic terms it was "no more than a colony, perhaps inferior to Australia and even to Canada," one deputy complained before the provincial assembly during the late Empire.[119] Developed largely as an economic subsidiary to the rest of the country, Rio Grande do Sul stood in a position of special weakness. An institutional framework molded to suit the export interests of the coffee planters of south-central Brazil could not work successfully for the southern frontier. At some point during the first half of the Old Republic, after a visit to Livramento on the frontier, Admiral José Carlos de Carvalho expressed this very well: "I verify the backwardness of the ranching of Rio Grande in comparison with that of the Plata, owing to the negligence of our leaders, who insist on administering the frontier state as if it were a central zone, without the competition of intelligent neighbors, who do everything to attract our commercial life to their ports."[120]

Marginal political status within Brazil slowed the modernization of the Campanha. As early as the 1820's, the liberal Gonçalves Chaves wrote of the damage that ill-conceived central policies could bring to the

Campanha and its capacity to compete. Yet a strong sense of grievance led nowhere as long as the regional oligarchies remained politically divided, and division characterized the complicated political history of the border region from the Paraguayan War throughout the remainder of the period under study. A civil war (1893–95) marked the early Old Republic in Rio Grande do Sul, and from it the PRR, with its focus on urban interests, emerged with rigid control of the state. It took the desperation of a further revolution (1923–26), heavily influenced by rancher economic disillusionment following the exuberance of the late phase of World War I, for the opposition groups to break the PRR's monopoly on power. Only after this cruel struggle did the political leaders begin cementing the regional oligarchies into a united front (the Frente Única). By the late 1920's, politicians were finally enacting policies in such areas as credit, freight rates, and tariffs that worked to the benefit of the Campanha in the face of the sustained competition offered by Rio Grande do Sul's "intelligent neighbors."[121] Riograndense rancher interests had finally found their political voice, one that became an influential underpinning in the struggle for the new form of Brazilian political economy that followed 1930.

CHAPTER EIGHT

The Campanha Circa 1920—and Beyond

B Y 1920 THE Campanha had witnessed great change. Meat prepared in the new Armour frigorífico began its journey by refrigerated railway car to markets as distant as the Mediterranean. On the
railway, journeys in the interior were no longer the adventure they had
seemed back in the 1850's, when travelers crossed the rivers in *pelotas*
(boats made of leather). The vistas remained impressive in their scope,
but now lines of wire fencing provided punctuation. Massive bulls of
impassive mien and bred from European stock were no longer a rarity.
The peon standing at the tick-bath seemed a world removed from the
slave-cowboy riding out to the rodeio and pondering a flight to freedom
across the southern frontier. Ignoring the absence of roads, rich estancieiros now took their families on picnics by car. Yet, behind the sure
drama of all such changes, the record of modernization was a mixed
one.

There were still some enormous properties in Rio Grande do Sul: 31
landowners declared areas greater than 20,000 hectares to the state
government in 1915.[1] However, the character of these huge estates was
much more diverse than it had been in 1850. Through astute (and even
ruthless) management, some members of traditional ranching families
had clung to vast amounts of land. For example, Januario Gonçalves
das Chagas was the third-largest landowner in the state in 1915, with
42,932 hectares of ranchland spread across five counties. Other great
tracts were controlled by relatively recent arrivals. The largest single
property (100,000 hectares on the Planalto) belonged to the Jewish
Colonization Association. Yet others lay in the hands of Rioplatense
salt-beef industrialists who had jumped a tariff barrier established under
the Old Republic. The Anglo-Argentinean George Clark Dickinson in
Itaqui (27,816 hectares) and the Uruguayan Emilio Calo in Quaraí
(24,451 hectares) were both in this category.

Even with the distorting influence of these major capitalists, most

Fig. 8.1. Prelude to a churrasco and dance on the Estância Progresso in Uruguaiana, c. 1916. From Monte Domecq et Cie., *Estado do Rio Grande do Sul,* p. 453.

counties now contained few properties whose dimensions resembled those of the sesmaria grants. The average sizes of the rural properties were now on a much smaller scale. Figure 8.2 provides clear expression of the separation of two distinct social and economic rural systems present in Rio Grande do Sul, the core of the ranching region in the southwestern zones of the state, where properties still averaged more than 400 hectares in size, and the colonial zone of the Serra, with properties under 100 hectares. Each of these emblematic systems had its outliers, most strikingly the isolated grassland zone of large properties on the northeastern border with Santa Catarina.

By 1920, ranchers talked of what they could do with their quadras (of 87 hectares each) rather than of leagues, of how to derive more from less. Families still arranged their compacts over land but the partible inheritance system had worn its influence into the decades.[2] Figure 8.3

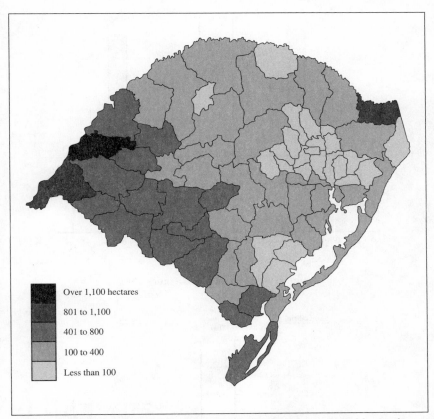

Fig. 8.2. The distribution of rural properties in Rio Grande do Sul by median size, 1920. Data from Brazil, *Recenseamento de 1920*, 3(2): 184–91.

shows the distribution of rural properties grouped by size category for four Campanha counties, including Alegrete, Bagé, and São Gabriel, those examined most closely at the opening of this study. In these three counties, the holdings of the greatest total areal extent were now in the range of 2,000 to 5,000 hectares each. There was a high degree of symmetry in the division of land, which was strikingly close when comparing Alegrete and Bagé. Discounting the smaller farms of little interest to ranchers (those with fewer than 100 hectares) and the giant properties (with more than 10,000), the average area occupied by landholdings in these three counties hovered around 1,000 hectares. While this was the general pattern for the Campanha, Figure 8.3 also includes the striking

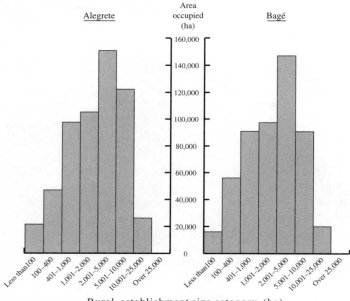

Alegrete

Bagé

Area occupied (ha)

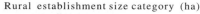

Rural establishment size category (ha)

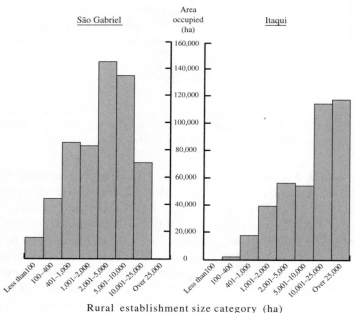

São Gabriel

Area occupied (ha)

Itaqui

Rural establishment size category (ha)

Fig. 8.3. The division of land in selected Campanha counties, 1920. Computed from data in Brazil, *Recenseamento de 1920*, 3(2): 184–89.

anomaly of Itaqui, where the distribution was heavily skewed toward large ranches. It showed an overall average rural property size of 4,019 hectares, a statistic that would not have looked out of place regarding the rawer frontier counties of Mato Grosso.[3] Old money reinforced by recent Anglo-Argentinean investment in land had shored up latifundismo in Itaqui.

Luso-Brazilian control over the Campanha ranchlands remained very strong, with the result that the overall direct landowning stake of foreigners was slender. The British held only sixteen rural properties in Rio Grande do Sul, but with an average area of 1,311 hectares, theirs were the largest of all those controlled by the foreign-born. Of greater significance were the 1,342 Uruguayans and their 703,095 hectares of land. With an average area of 524 hectares, these holdings were presumably either ranches or large commercial farms. Much of this land had probably belonged to gaúchos repelled from the state by their inability to meet the costs of modernization.[4]

The change in style in the promotion of pastoral interests must have seemed remarkable to some of those born under the Empire. For example, during 1919, Ildefonso Simões Lopes, the minister of agriculture, had shown a film in Rio on the ranching and agriculture of Rio Grande do Sul to the president and the other ministers.[5] A newspaper report carried the encouraging message back into the south that "each pure breeding animal (*reproductor*) that appeared was greeted with admiration."[6]

While federal politicians had progress in ranching delivered to them on film, others boarded train or ship and went to look for themselves. Toward the end of 1920, a Porto Alegre newspaper published a summary of ranching in Rio Grande do Sul prepared by a party of visitors from Argentina. This delegation spoke of grazing lands with unpalatable grasses and of the hindrance to breeding occasioned by the continuing prevalence of the carrapato. But its members also held that the grazing zones of Rio Grande do Sul were becoming more refined; they saw the same process taking place as had happened "with the formerly unpalatable pastures of Buenos Aires, Entre-Rios, Santa Fé, and in the Pampa [which were] today splendid alfalfa pastures or lands of the first class." On the issue of cattle breeding, the delegation proved equally optimistic, repeating the cliché that this was largely a question of time and money. They described the interest that the estancieiros of Rio Grande do Sul had shown in the Zebu, an animal the delegates characterized as one of "ugly aspect and tough meats," as a thing of the past. The local

Fig. 8.4. Purebred Blackface and Lincoln sheep on the Estância da Bôa Vista in Porto Alegre, c. 1913. From *A Estância* (Apr. 1913): p. 34.

ranchers were now fully engaged in crossbreeding the cows of the region with high-quality bulls of European breeds.[7]

Sheep had also finally become a serious business on some of these vast grasslands. In contrast to the pattern in the Río de la Plata, cattle still outnumbered them in Rio Grande do Sul (8,489,496 cattle and 4,485,546 sheep).[8] The proportion of the properties carrying sheep declined with distance from the Uruguayan frontier (Figure 8.5). Following the Uruguay River northward, for example, while 73 percent of landholdings in Uruguaiana carried sheep, this fell to 32 percent in São Borja.[9] That sheep were mainly associated with large properties can be seen by the very high proportion (86 percent) of properties carrying them in Itaqui.

Signs of the emergence of Rio Grande do Sul as a larder for the whole of Brazil were already evident in a string of national firsts.[10] Rio Grande do Sul produced almost all of the country's wheat.[11] It contained the counties producing the most maize and tobacco.[12] It headed the lists for the numbers of plows and tractors in the country.[13] However, by 1920 land-use diversification in the Campanha had still played only a very

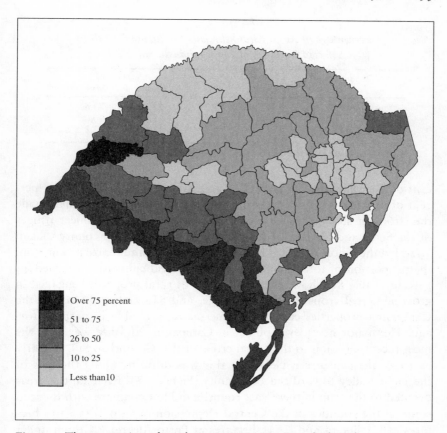

Fig. 8.5. The proportion of rural properties in Rio Grande do Sul carrying sheep, 1920. Calculated from data in Brazil, *Recenseamento de 1920*, 3(1): 486–91.

slender role in the emergence of this commercial agriculture. Most of the agricultural dynamism was to be found in the colonial zones of the northern half of the state. The contribution of areas of very recent settlement was striking, especially those parts of the Planalto served by the Santa Maria–São Paulo railway. The railway from Santa Maria across the Planalto had reached the forests of Paraná pine around Erechim in 1910; an exceptionally rapid agricultural development followed its arrival. A decade later, the area planted with cereals in Erechim already far outstripped that in any Campanha county.[14]

In 1920, all of the Campanha counties still cultivated less than 1 per-

TABLE 8.1

*Percentage of Rural Establishments Producing Cereals in
Selected Counties of Rio Grande do Sul, 1920*

	Total holdings	Rice	Wheat	Maize
Alegrete	1,118	2.2	22.0	73.6
Bagé	904	0.8	18.1	80.0
São Gabriel	921	5.1	12.8	72.8

SOURCE: Computed from data in Brazil, *Recenseamento de 1920*, 3(2): 86–90.

cent of their territory.[15] The land in cereals made up as little as 0.1 percent of the total area of Uruguaiana, the county where sheep had made
the strongest showing. This lay in stark contrast to the pioneer fringes
of the Serra and Planalto. For example, in Guaporé, the county that in
1920 led the whole of Brazil in quantity of wheat and maize production,
the proportion of total county area under cereal cultivation reached 43
percent. Table 8.1 shows the proportions of rural properties engaged in
growing cereal crops in Alegrete, Bagé, and São Gabriel. Most of the
Campanha properties cultivated some maize, probably in part for forage. The major new development in Campanha land use was rice, the
most recent addition to the cereal crops of Rio Grande do Sul. As early
as 1920, the evidence is clear that this was diffusing along the axis of
the Jacuí Valley toward the Campanha (Figure 8.6). Even so, the areas
devoted to the crop in these vast counties did not compare with those in
some of the counties of the Central Depression or Litoral (11,298 hectares at Cachoeira and 6,945 hectares at Porto Alegre).[16] Grown under
irrigation, rice was an expensive crop to cultivate. Apart from water
availability, the leading factor explaining the spatial distribution of rice
in this early phase was the proximity of commercial capital. In this regard, it seems appropriate that the Granja Cascalho, one of the most
highly capitalized farms in Brazil (the property of Pedro Osório, the
"rice king"), was located on the site of a former charqueada in the outskirts of Pelotas.

Both the census and contemporary photographs attest that tractors
were in use on the large farms of major ranching counties by 1920. Although the amount of commercial cultivation was limited, the degree of
mechanization had already reached high levels in some more prosperous
ranching counties, such as Alegrete and Bagé. The advertising campaigns of companies like Sulford of Porto Alegre, which drew attention

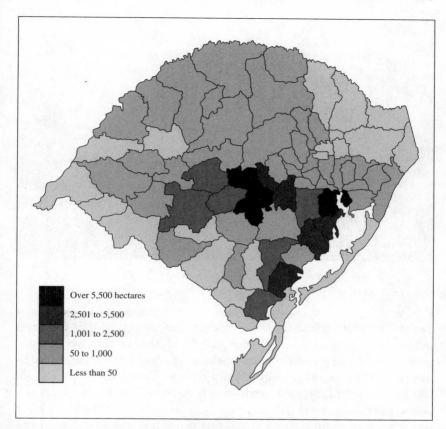

Fig. 8.6. The area of Rio Grande do Sul cultivated in rice, 1920. Data from Brazil, *Recenseamento de 1920*, 3(3): 176–78.

to the success of Fordson tractors in the British government trials of World War I, were working successfully.[17] For example, with 25 tractors, Alegrete had one machine for each 117 hectares of its cultivated land. This might have seemed impressive to the estancieiros of São Borja, where there were still only two tractors. Yet all the Campanha counties together were eclipsed by Erechim with 311. In 1920, the dynamic agriculture of that pioneer county accounted for almost a fifth of all the tractors in Brazil.[18]

With few exceptions, ranching and cultivation remained separate agricultural systems in the Rio Grande do Sul of 1920. That Bagé pro-

Fig. 8.7. Preparing for rice cultivation: a Hart-Parr tractor on the Granja Cascalho in Pelotas, c. 1915. From *A Estância* (July 1915): 622.

duced a little under 900 tons of wheat represented agricultural progress, but of a most limited kind, even compared with the Serra county of Guaporé, let alone Argentina. Those ranchers who had heeded the speeches imploring them to hasten cultivation still plowed lonely furrows. The attention of most of the estancieiros was firmly fixed on the improvement of their ranching. Generally copying Argentina, which set the regional standard, their main preoccupation was producing animals to meet the demanding requirements of the frigoríficos, so they were trying to find the money to finance crossbreeding and internal fences. In a region where planted pasture had never taken hold, Campanha ranchers worked at intensification for the most part on natural pasture. Figure 8.8 shows the density of cattle and sheep (1 head of cattle = 4 sheep) per hectare of pasture in 1920.[19] The data show clearly the results of intensification around the cities of the Litoral, where there were already incipient dairy belts. High densities of livestock in the Serra were linked with the stall feeding of cattle on small properties. But in the ranching zones, the pattern in Figure 8.8 bears out what contemporary observers emphasized: cattle and sheep had reached their highest stocking levels near the Uruguayan frontier, notably on the campo fino in the southwestern part of the state, where densities compared with those on the pampas of Argentina.

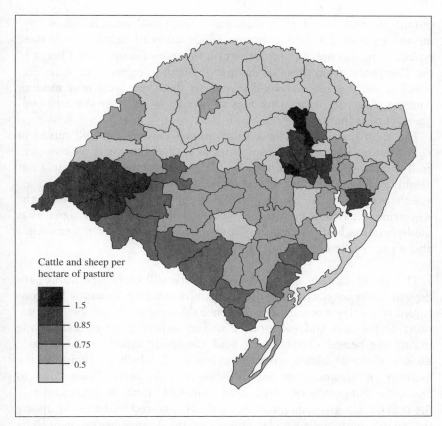

Fig. 8.8. Livestock density distribution in Rio Grande do Sul, 1920. Based on data in Brazil, *Recenseamento de 1920*, 3(1): 415–17 and 3(2): 111–13, 312–17.

The contemporary qualitative assessments of Campanha ranching were equally mixed. Leo Plant, a representative for Wilson, the American packing company, asserted that the state of breeding in Argentina and Uruguay had gone "beyond the horizon" in comparison with that in Brazil, whose exported meat could find sale only to armies rather than the civilian populations of large European cities.[20] Commenting on the price of Brazilian frozen meat at Smithfield, the London *Evening Standard* noted that it was quoted for just a penny less than the Argentinean product, "when this is fit for a king, while the Brazilian is only fit to feed the lions at the zoo."[21] The reputation of the Campanha suffered unfairly in such broad generalizations. Europeans, uninterested in buying the meat of the Zebu cattle on which many of Brazil's herds were based, were also unready to distinguish the more Europeanized herds of the Campanha from those of the rest of Brazil in terms of their reputation.[22] If the progressives could take heart that there were now modern packing plants in place, there was still much work to be done to conquer overseas markets.

The record of Riograndense ranch modernization was still mixed in 1920. But aware that their readership in greater Buenos Aires was generally uninformed about conditions in the borderlands of their giant neighbor, the visiting Argentinean delegation had gone so far as to argue that there was "a danger in the south of Brazil for the wealth of the Argentinean pastoral industry."[23] There could be no clearer signal that modernization had reached the Campanha than worry in Buenos Aires about potential competition.

The broad changes analyzed in this work still carry significance today. In the years since 1920, the Campanha ranching economy has continued to display a remarkable resilience throughout the various fluctuations of political and economic life. The technological changes made during the period of this study held essentially intact until the early 1960's. Planted pasture remained exceptional, which set limits on the quantity of livestock that the Campanha could carry. From 1920 to 1960, the Riograndense cattle herd remained stable at approximately 8.5 million. Crossbreeding continued slowly and received a boost during the 1940's with the financial policies of the Vargas government. The number of cabanhas (highly capitalized specialist breeders) multiplied, especially during the 1940's. The increasing tendency to keep sheep as well as cattle, already evident by 1920, went much further; from 1920 to 1960, the Riograndense flock tripled in size.[24]

There are clear elements of continuity. Most of the Campanha is still in natural pasture owned by Luso-Brazilians. In 1980, only 6.1 percent of the total pasture area was planted.[25] The landholdings of the Campanha remain strikingly large in comparison with those of other subregions of Rio Grande do Sul, their median size exceeding the state average by a factor of four or five.[26] Although reduced in size since 1850 through the creation of new counties, the modern county of Alegrete still contains the largest single cattle herd in the region, greatly changed in quality since the mid-nineteenth century, but not much in number.[27] Ranchers are still debating the relative merits of specific breeds. The debate of the early part of the century on European breeds versus the Zebu has recently resurfaced.

There are also elements of change. The pace of technological improvement in Campanha ranching has quickened since about 1960. Today Campanha ranchers are part of an international network with access to current research about genetic improvements in breeding, as well as other aspects of herd and flock management. In 1990, Bagé ranchers bid on sheep bred from New Zealand's supreme champion ram. While ranchers promote sales of Suffolk and Ile de France sheep in the Assis Brasil exhibition park outside Porto Alegre, many breed names no longer reflect specific geographical regions.[28]

As during the period under study, the uncertain course of Brazil's political economy continues to buffet Campanha ranchers. However, the best placed are streamlining their operations and adopting new technology on an ever-greater scale. The best cabanhas sell their animals not only to other parts of Brazil; their sheep are also sold to Argentinean and Uruguayan breeders, a reversal of the pattern common before 1920. They combine large-scale rice cultivation and specialized breeding of pedigree animals with the meat and wool production of 1920.[29]

The trend is for a steady decline of the area in pasture. At last, the fusion of two major agricultural systems is beginning to occur—but very slowly. As of yet, the key crop is rice, much of it grown by tenant farmers on estância land as part of rotational systems designed to improve ranch pastures. Rice cultivation is thus the leading means of ranch intensification in the Campanha. There are broad parallels in this with the wheat/alfalfa rotations of Argentina a century ago. The 236,502 hectares planted with rice in 1980 in the Campanha are a far cry from the 2,415 hectares of 1920.[30] Today rice is the leading product by value coming from the southern part of the state, at least as far as official statistics are concerned (according to some authorities, much cattle slaughter remains clandestine).[31] Accounting for over 40 percent of national rice production, Rio Grande do Sul continues its already long history as a food producer for other parts of Brazil.[32]

Since the 1970's, the evidence has pointed to potential high returns from cultivation with careful management throughout the grasslands of southern Brazil. Gerd Kohlhepp has pointed to some of the recent successes of innovations on the grasslands of Paraná. Genetic research is leading to higher-yielding varieties of crops that pay heed to local environmental conditions. In the direct drilling of crops to reduce soil erosion risks, southern Brazil is second only to the United States in terms of area. In general, the search has intensified for rotations of summer and winter crops that are in equilibrium, thus leading to ecologically sustainable systems.[33] The twenty-first century may well see an end to what Leo Waibel, in the late 1940's, called the "cancer" of the economic and spatial separation of cultivation and ranching.[34] But the integration of these land-use systems also involves social problems, which will call for much human ingenuity in their resolution. The greatest challenge is to achieve a balanced development with the minimum waste of human capital. In this, the past surely carries lessons.

Conclusion

THE TABLE NEEDS of industrializing Europe (especially Britain) engendered stimuli for modernization that extended to the realms of ranch production and processing in southern South America. Frigoríficos called for more flesh on the bone, hence breeding; highbred animals were more demanding in their forage habits, hence planted pasture. Given these linkages, we might theorize that a package of technical elements of modernization should have appeared at roughly the same time on the ranch as in the processing of its products. In practice, the geographical patterns of innovation were much more complex. The path to the improved estancia, where pedigree livestock ate planted forage in preparation for a long journey to a specialized market, did not follow an even course, either in time or in space.

In its patterns of modernization, the southern South American ranching region exhibited a clear core-periphery structure. The empirical evidence presented in this study exemplifies the broad outline of Friedmann's theory of polarized development.[1] Innovations came later to Rio Grande do Sul, part of the periphery of the temperate grasslands, than to Buenos Aires Province and to southern Uruguay. Selected livestock, wire fencing, railways, and even American-owned frigoríficos—to take a few of the key innovations for the intensification of ranch production—all diffused to the Campanha from Uruguay and Argentina, especially the former. By its earlier innovation, the core of the southern South American ranching region managed to dominate the periphery; for example, profits from the sale of breeding stock and forage reinforced growth in the core. Yet the grasslands of Rio Grande do Sul did not provide a mirror image of the rural experiences of Argentina and Uruguay. Physical and human elements combined in a distinctive geography of economic change in the region. While economic affairs in the Campanha of Rio Grande do Sul were very tightly connected with those in the Plata, the course of its ranching modernization was different.

A significant early difference from the Plata was that Rio Grande do Sul missed the wool boom of the 1850's and 1860's, described by Barrán and Nahum as probably the most radical transformation in the whole of Uruguay's rural history.[2] Wool provided a higher income per unit area than cattle, providing smaller properties with an economic rationale they had previously lacked. Besides its economic value, wool had the power to change society and politics. It attracted immigrants from new sources. It provided a strong early impetus for fencing. Without this stimulus, the Campanha was already a couple of decades behind the Plata in innovation even before the Paraguayan War.

And in some respects the time lag was even longer. After the midnineteenth century, with the arrival of European immigrants on a large scale, the social structure of the Río de la Plata countries began to diverge sharply from that of Brazil. In Rio Grande do Sul, slavery was still central to the operation of the charqueadas and even to some ranches as late as the 1880's. While Rio Grande do Sul had its visionaries in the imperial period, they were few, and their capacity to realize change was limited. Unlike specific zones of the Río de la Plata, usually the most temperate ones, the Campanha failed to attract immigrants keen to diversify land use and ready to challenge the inherited values of a traditional society.

Much has been written about Brazil forming part of Britain's "informal" empire, and there is evidence to support this case. Yet, away from the Rio de Janeiro–São Paulo axis and the major cities of the Río de la Plata estuary, British influence in southern South America displayed elements of distance decay. In Rio Grande do Sul, it failed to move much beyond the port of Rio Grande. By the 1860's, as political hindrances to commerce in the Plata disappeared, merchants began to serve Rio Grande do Sul from Montevideo, a rational decision given that city's location. Montevideo stood "at the entrance to a powerful hydrographic system which drained the most densely populated part of South America in the nineteenth century."[3] Throughout the second half of the nineteenth century, British influence in Rio Grande do Sul steadily weakened. The direct northwest European contribution to the modernization of ranching was minimal there, in marked contrast to its southern neighbors, where British capital and initiative moved beyond the cities far more. In Rio Grande do Sul, the domestic mercantile elite and foreign-born merchants perceived themselves as holding separate interests. There is little evidence here of the powerful linkages between elites that are central to dependency interpretations. Some might argue that the absence of this melding of elite interests saved Rio Grande do Sul from the ultimately unbalanced growth of "unequal exchange" on a massive scale with Britain. But this fractioning of the elite also clearly slowed the transition from mercantile to industrial capital in the processing of ranch products. Refrigeration technology involved the assembly of massive pools of capital that far outstripped the means of individual families within the mercantile elite.

As well as standing at the periphery of the functional region formed of the temperate grasslands of southern South America, Rio Grande do Sul was also on the margin of a formal region, the huge nation of Brazil. Political marginality within the imperial system was one of the most significant constraints on Rio Grande do Sul's development. The Brazilian Empire neither challenged nor equipped the latifundia of the Campanha. With the threats of land reform or even land taxes distant in imperial Brazil, ranchers appreciated that there was an economic rationale for extending the scale of the private domain as far as possible. The "archaic" mentality, which Barrán and Nahum see as the bedrock of rural Uruguay, was even more deeply entrenched in the conservative rural society of the Campanha. Without credit, tariff protection, or even legal reforms to encourage fencing, "tradition" dissolved more slowly than in the Plata.

The Brazilian Empire's failure to provide infrastructure has been identified as a key factor in the country's generally poor economic performance before 1890.[4] As Leff has argued persuasively, this was an important national issue, but its seriousness had a varied regional expression. Although the "let the cattle walk" mind-set of the cynics who spoke against early provision of infrastructure to the sparsely settled Campanha is readily understood, the failure to provide infrastructure on Brazil's temperate fringe was an especially serious omission. By the 1890's, Uruguay had one of the densest railway networks in South America. Only when the Campanha looked set to develop as an economic appendage to a newly consolidated Uruguayan nation-state did public policies emerge from Rio de Janeiro that cleared some of the obstacles to Riograndense development. Beginning in the late 1880's, abolition of slavery, railways, and tariffs laid the groundwork for a slow transformation of Campanha ranching.

Challenging the contemporary core regions around Buenos Aires and Rio de Janeiro, as early as 1815 Uruguay's great caudillo José Artigas promoted the economic integration of Uruguay, Rio Grande do Sul, and Mesopotamian Argentina, based partly on an awareness of the natural geographical linkages among them.[5] This notion is alive today in the form of MERCOSUR, the recently established common market of the southern South American nations, with its administrative secretariat located symbolically at Montevideo. Leff has argued that Brazil's classic "problem region," the northeast, might have experienced more favorable rates of economic growth had it been a separate entity from the Empire, one able to design its own fiscal policies.[6] The northeast was clearly not the only "victim" of a political economy designed to suit the coffee planters of south-central Brazil.

During the period under study, ranching and colonial farming remained strikingly separate in Rio Grande do Sul. The Empire did far more to promote the extension of the latter than the intensification of the former. While Brazil scoured Europe seeking immigrants to colonize and farm the pioneer lands of the Serra, Brazilians not many hundreds of kilometers away were being cleared from the ranches after 1880 with no clue as to where they would turn next.[7] Unlike the pampas of Argentina, the Campanha could not be transformed through tenant farming.[8] Environmental obstacles stood in the way of the plow; for example, unlike Buenos Aires Province, most of the Campanha was not suitable for the large-scale cultivation of either wheat or alfalfa, something most ranchers seem to have understood better than their few agronomists, and even some modern historians. In addition, before the middle 1880's, the continuing presence of slaves on some of the ranches posed a major obstacle to European immigration on any significant scale. Sheep, cultivation, immigrants, and foreign mercantile capital all presented weaker challenges to tradition in the Campanha than in the Río de la Plata, slowing the economic development of Brazil's southern frontier. There is little in Rio Grande do Sul to compare with Scobie's "revolution" on the wheat frontier of the pampas of Argentina.

John Lynch has argued that an economy based on hides and salt-beef "could not generate growth," based on his study of Rosas and the Argentina of the first half of the nineteenth century. Lynch's conclusion does not hold for Rio Grande do Sul.[9] Even while based predominantly on traditional products, technical refinement raised output from the Campanha ranching staple economy. Many of the innovations described in Chapter 5 were adopted under the stimulus of the domestic as well as the international markets. Given the weaker pull of the international markets in the Campanha than in the Plata, the slowness of change in the former can be more readily understood. Campanha ranchers worked with a physical environment that included more risk in experimentation (with breeds and grasses, for example) than did their counterparts on the humid pampa. The weakness of Rio Grande do Sul's infrastructure, especially of its major port, meant that the cost of capital was higher, and the profits to be gained from adopting more productive technology less certain, than in the regions to the south.[10] Many Campanha ranchers appeared to fall short of the liberal ideal, like their late-twentieth-century descendants more given to satisficing than to optimizing.[11] Yet after 1880, despite the complex variety of constraints the ranchers faced over the period analyzed, the ranches were slowly transformed. Sandra Pesavento's study of the slow modernization during the Old Republic, written as an analysis of dependent capitalism within the periphery, rests on sure ground in showing how ranchers were usually starved of capital. But her thesis that Riograndense ranching was exhibiting stagnation on the eve of World War I does not fit with the patterns of change established in this study.[12] After 1895, there was general, if gradual, improvement to the ranches; the pace of change quickened after 1914 as ranchers responded to the newly available credit that was part of the extraordinary stimulus of World War I. For confirmation that things were changing we need look no further than to the iconography of advertising in the specialized press.

This study ends with the arrival of the leading American companies in Rio Grande do Sul. By the early 1920's, major fixed investments in packing plants had already been made, affirming the modernity of the ranching economy. In 1920, the fuller use of the frigoríficos awaited the further refinement of the herds, as well as the political settlement of the postwar strangulation of cheap credit to ranchers and of the continuing high freight rates within the regional transport network. In the resolution of these issues, Rio Grande do Sul benefited greatly from the administration of Getúlio Vargas (1928–30). As leader of the revolution that brought down Brazil's Old Republic in 1930, Vargas took Rio Grande do Sul to the forefront of national attention.

The changed political economy of Brazil in the period since 1930 has been accompanied by a remarkable dynamism. The economic nationalism of the 1930's marked the beginning of Brazil's extensive diversification of its economy and of the long climb toward its position as one of the leading economies of the developing world. In recent decades, Brazil's trading relationships with its southern neighbors have changed dramatically. Today the polarization effects in the ranching zones of southern South America are of a completely different character than during the period under study, testimony to the importance of political economy in development. In recent years, the price differential for land on either side of Rio Grande do Sul's frontier with Uruguay has been huge, with land between three and six times more expensive in Brazil, the reverse of the pattern at the end of this study.[13] The disparity is linked in large part to Brazilian success with rice cultivation.[14] Brazil's long tradition of pushing southward is still with us.

This work has paid considerable attention to the character of traditional society as an essential prelude to any analysis of the nature and impact of modernization. The "traditional" ranching culture of Rio Grande do Sul was not an inert stage on which the drama of modernization unfolded. This study does not stand alone in arguing the key significance of the nineteenth century for our understanding of the geographical character of the Campanha. In recent years, there has been an enormous emphasis on "traditional" gaúcho culture through, for example, rodeios and native song festivals. By idealizing the estância as the concrete and spatial representation of a supposedly egalitarian and harmonious society, some social scientists have argued, the state has equipped itself with a powerful ideology designed to keep submissive the marginal elements of a now mainly urban population. And they have further argued that the official cult of the estância as the key geographical symbol of Rio Grande do Sul supports the current resistance of some of the estancieiros to change.[15] Such symbolic interpretations of the past are highly selective. While the gaps between landowners and labor widened after 1850, the Campanha at mid-nineteenth century was no egalitarian place. The evidence on ranch slavery in Chapter 2 alone confirms this argument.

There were marked regional differences in the timing and character of how the southern South American ranching economies were drawn into the international economy. Comparisons between the single region of the Campanha and the national experiences of Argentina and Uruguay are crude exercises. Much more regional research is required, not least on Argentina's grasslands beyond Buenos Aires Province and in what two scholars have recently termed the "difficult terrain of the history of ecosystems and agricultural technology."[16] This study is offered in the hope of encouraging others to work toward the elaboration of a more sophisticated historical geography of regional change in southern South America.[17]

Given that the Latin American economies became so outward looking during the period under study, it is understandable that research has concentrated on the relationships between core and periphery at the international scale. But polarized growth was not confined to relations between Britain and Argentina or Britain and Brazil. Research on resource transfers within the periphery is still required. Harold Brookfield has argued that development geography ought to stress innovation "arising at points in a system."[18] In this regard, the patterns of northwest European entrepreneurship in the Río de la Plata have still to be addressed systematically; this will undoubtedly be an important study. A major study of Brazilian involvement in Uruguay during the nineteenth century is also needed. As Argentina, Brazil, Paraguay, and Uruguay plan the integration of their now-disparate economies, the study of interaction across the frontiers in historical perspective takes on a timely significance.

This study has elaborated change at what is still a broad scale. Consuls, provincial presidents, and even county councillors usually generalized about vast pieces of territory in their writings. Only a few of the sources examined have permitted conclusions about interranch variations. For this, case studies of individual counties and even sections of counties are still needed; both for ranching and for colonial farming, Rio Grande do Sul still awaits its *Vassouras* (Stanley Stein's classic study of a Rio de Janeiro coffee county).[19] The distribution maps of innovations presented in this study help in the identification of likely candidates for county-scale studies of ranching. While it may be optimistic to assume the survival of extensive estância records, there are worthwhile studies to be made at the microscale, using notarial documents.

Within a vast and now predominantly urban Brazil, the Campanha remains a very singular landscape. Despite the great transport improvements of recent decades, there are still vestiges in these rolling hills of the isolation described so well by nineteenth-century authors. And in the winter, the Minuano wind still rips across the grasslands. Ranch hands taking their *mate* tea round a fire, their horses tethered to the fences, reflect a sense of permanence and tranquility. A traveler who follows the crushed-stone roads across the coxilhas will recognize aspects of the mid-nineteenth-century "genre de vie."

Reference Matter

Abbreviations

AAHRGS	*Anais do Arquivo Histórico do Rio Grande do Sul*
AALRGS	*Anais da Assembléia Legislativa do Rio Grande do Sul*
AHI	Arquivo Histórico do Itamarati
AHRGS	Arquivo Histórico do Rio Grande do Sul
APRGS	Arquivo Público do Estado do Rio Grande do Sul
BN	Biblioteca Nacional (Rio de Janeiro)
CHLA	*The Cambridge History of Latin America*
CM	Câmara municipal
CV	Coleção Varela (Alfredo Varela collection of documents), AHRGS
H. of C.	House of Commons (London) sessional papers (Parliamentary Papers)
HRUM	*Historia rural del Uruguay moderno*
MHN	Museo Histórico Nacional (Montevideo)
N, M, E	*número* (number of inventory), *maço* (packet), *estante* (shelf)
P. of P.	President of province
RAF	*Revista Agrícola da Fronteira*
RARGS	*Revista Agrícola do Rio Grande do Sul*
RARU	*Revista de la Asociación Rural del Uruguay*
RGS	Rio Grande do Sul
RIHGRGS	*Revista do Instituto Histórico e Geográfico do Rio Grande do Sul*
V.S.L.	"Informe do visconde de São Leopoldo, sem indicação de destinatário, sobre a criação de gado na Província do Rio Grande do Sul" (Rio de Janeiro, 1842), manuscripts, II-36, 1, 18, BN.

Notes

Introduction

1. In recent years, the controversy has come mainly from the Amazon basin, where ranching as a leading cause of deforestation became a significant international public policy issue. There is an increasingly rich literature on this topic; see, for example, Hecht, "Environment, Development and Politics."

2. Born in São Borja, Getúlio Dorneles Vargas was Brazil's longest-serving and most influential chief of state. He presided over authoritarian regimes from 1930 to 1945 and was the constitutional president of Brazil from 1951 until his suicide in 1954. For accounts of his career, see Dulles, *Vargas of Brazil*; Bourne, *Getulio Vargas*.

3. For a basic introduction to the physical geography of the region, see James and Minkel, *Latin America*, pp. 453, 469.

4. To what extent the gaúcho of Rio Grande do Sul differed in such matters as origins, costumes, and habits from the gaucho in the Plata remains a matter of debate. See, for example, the short discussion in Love, *Rio Grande do Sul and Brazilian Regionalism*, pp. 3–4. Originally a pejorative term for those who lived beyond the law, the label "gaúcho" was more generally applied to those engaged in ranch labor by the period of this study. Drawing on a romantic appreciation of the pastoral complex and the regional warrior heritage, today all the inhabitants of Rio Grande do Sul are known as gaúchos. Even in the highly urbanized modern state, interest in the pastoral culture is intense, and with its estimate of over two million active members, the Movimento Tradicionalista Gaúcho claims to be "the largest popular culture movement in the Western World." See Oliven, "Growth of Regional Culture in Brazil," p. 110.

5. Hennessy, *Frontier in Latin American History*, p. 86.

6. Galloway, "Brazil," p. 371.

7. See David Robinson's comments in his thematic survey of the broader region's historical geography ("Historical Geography in Latin America," pp. 181–82).

8. See Friedmann, *Urbanization, Planning, and National Development*, especially pp. 41–64.

9. The most energetic efforts to "civilize" life in the Campanha were made by Joaquim Francisco de Assis Brasil (a distinguished politician and lobbyist for rural matters) within the castle-like house of the Granja de Pedras Altas. His career has been the focus of increasing research attention of late. See, for example,

Reverbel, *Pedras Altas*; Brossard, ed., *Idéias políticas de Assis Brasil*, which ranges beyond politics despite the title. For the perspective of a child growing up in the special environment of Pedras Altas, where one day might bring a reading of Rudyard Kipling and the next a viewing of newly arrived imported Devon cattle, see Reverbel, ed., *Diário de Cecília de Assis Brasil*, p. 12.

10. Richard Graham, *Britain and the Onset of Modernization in Brazil*, p. 15.

11. Cardoso, "Rio Grande do Sul e Santa Catarina," pp. 481–82. See also Cardoso, *Capitalismo e escravidão no Brasil meridional*, especially p. 45. This book, which developed out of a doctoral thesis, is of seminal value for understanding the development of the economy and society of Rio Grande do Sul.

12. In Argentina, for example, even for the early part of the twentieth century, Peter Smith has demonstrated how "the ranchers had extremely easy access to the decision-making centers of government" (*Politics and Beef in Argentina*, p. 247).

13. See Friedmann, *Regional Development Policy*, p. 15.

14. See Langer, *Economic Change and Rural Resistance in Southern Bolivia*, pp. 1–2. There is an excellent review of dependency theory in the first three chapters of Abel and Lewis, eds., *Latin America, Economic Imperialism and the State*. See, especially, the "General Introduction," pp. 1–25. See also Kay, *Latin American Theories of Development and Underdevelopment*. For a broad view of dependency in Brazil that concentrates on the present but also pays attention to the structural roots of the country's modern economy, see Evans, *Dependent Development*, especially pp. 55–100.

15. See Mosk, "Latin America and the World Economy."

16. Grigg, *Agricultural Systems of the World*, p. 241.

17. Hobsbawm, *Industry and Empire*, p. 112.

18. On this theme, see Blainey, *Tyranny of Distance*.

19. Hobsbawm, *Industry and Empire*, p. 122.

20. Gootenberg, "Beleaguered Liberals," p. 64.

21. Stein, *Vassouras*, p. vii.

22. On the stirrings of modernization in Brazil during the 1850's, see Richard Graham, *Britain and the Onset of Modernization in Brazil*, especially pp. 23–27.

23. Barman, *Brazil*.

24. As the governor of Buenos Aires between 1829 and 1852, Rosas dominated the politics of Argentina in the first half of the nineteenth century. His regime left an indelible mark on the agrarian structure of that country. Domingo Faustino Sarmiento, the famous liberal statesman, explained the Rosas system as "a monstrous product of the estancia" and complained that everything was influenced by the interests of the major rancher caudillos: "What are Rosas, Quiroga and Urquiza? Cowboys, nothing more." See Lynch, *Argentine Dictator*, pp. 1, 89.

25. CM of Alegrete, doc. 936, 10 Jan. 1858, AHRGS.

26. Vidal de la Blache, "Les genres de vie dans la géographie humaine."

27. Feder, *Rape of the Peasantry*, p. 79.

28. Government of Brazil, *Recenseamento de 1920*. Most of the data on ag-

riculture can be found in volume 3. The significance of this census is greatly heightened by the fact that Brazil did not take another until 1940.

29. Cooper, *The Brazilians and Their Country*, p. 226.

30. Waibel, "Regiões pastoris do hemisfério sul," p. 61. This article, Waibel's first in economic geography, was published originally in German. Looking at the improvements made to the Brazilian cattle industry during and after World War I, the American geographer Clarence Jones forecast a bright future for it, if sufficient improvements were made in the areas of animal quality, elimination of disease, and transport ("Evolution of Brazilian Commerce," p. 562). For further contemporary expressions of optimism about the Brazilian meat industry, see Downes, "Seeds of Influence," especially pp. 467–79.

31. On the impressive modernization of Argentina's economy and society during the latter half of the nineteenth century and the opening years of the twentieth, see, especially, Scobie, *Revolution on the Pampas*. For the equally striking socioeconomic transformation of Uruguay, see Barrán and Nahum, *HRUM*; see also their *Batlle, los estancieros y el Imperio Británico*.

32. Even in the late 1860's, as Argentina was only beginning to experience its rural transformation, there were already considerable differences in the prices paid for land intended for sheep breeding in the different regions of the Río de la Plata. These were partly a reflection of the degree to which the natural pastures had already been modified. See Latham, *States of the River Plate*, pp. 178–82. A Brazilian journalist who analyzed the transformation of Argentina around 1880 was quick to point out that the country's wealth was largely confined to the province of Buenos Aires and a part of Santa Fe. *A Pátria*, 13 Jan. 1881 (article reprinted from the *Gazeta de Porto Alegre*).

33. The scope for comparative work on cattle frontiers is considerable. For a brief introduction to their nature in Latin America, see Hennessy, *Frontier in Latin American History*, pp. 82–89. A limited but useful introduction to the comparative study of southern South American ranching is Slatta, "Gaúcho and Gaucho." Commercial agriculture on the southern South American grasslands has been placed in a stronger comparative frame than ranching; see especially the recent studies by Solberg, *The Prairies and the Pampas*, and Adelman, *Frontier Development*.

34. Warwick Armstrong has written about the comparative development of Argentina, Australia, and Canada (privileged periphery societies) in the period 1870 to 1930. The argument advanced in this paragraph is in full agreement with Armstrong's claim that "the primary basis for understanding the nature of a society's evolution and its potential for change must centre on its current internal situation set in the context of its historical experience." For a brief overview of his ideas about the importance of research on the social structures and relationships within peripheral societies, see Armstrong and McGee, *Theatres of Accumulation*, pp. 34–37.

35. For the argument that Britain held an "informal empire" around the Río de la Plata during the nineteenth century, see Winn, "British Informal Empire in Uruguay in the Nineteenth Century." See also Ferns, "Britain's Informal Empire in Argentina," and *Britain and Argentina in the Nineteenth Century*. Given the extensive German emigration to southern Brazil that took place during the

nineteenth century, the attitudes of German governments toward these regions are interesting. On competition within British and German methods of "informal imperialism" in southern Brazil during the decades before World War I, see Forbes, "German Informal Imperialism," pp. 385–86, 398.

36. Winn, "British Informal Empire in Uruguay in the Nineteenth Century."

37. See the discussion of Graham in Bell, "Foreign Investment and the Historical Geography of Brazil," pp. 37–38.

38. For a short summary of their views, see Barrán and Nahum, *HRUM*, 1(1): 199–200. The Brazilian author Guilhermino Cesar has drawn attention to the interdependency between Rio Grande do Sul and Uruguay during the early nineteenth century ("Ocupação e diferenciação do espaço," p. 22).

39. W. Gifford Palgrave, a British consular official, gave a succinct description of the "loser" mechanics within a generally optimistic report on Uruguay made in the 1880's: "I regret that from this favourable notice I must, speaking broadly, deduct the grazing grounds held by Brazilian owners, who, to the great disadvantage of the Uruguayan territory, are proprietors of at least a fourth, perhaps a third, part of its pasture-lands. These Brazilian land-owners, often absentees in person, and almost invariably so, so far as their savings are concerned, abstain for motives into which it is not worthwhile to inquire, from anything like improvement in the quality of their stock; as also from a proper use of the ground itself, its sub-division into mutually successive pasture-grounds, and the husbanding of its water-supply . . . a practice universal among the landed proprietors of Uruguayan or European nationality, but almost unknown on Brazilian-held pastures, while at the same time the Brazilian landlords, acting what is commonly called a 'dog in manger' part, by demanding the most exorbitant prices on occasion, preclude the buying up of their vast but comparatively useless holdings by customers who could and would make a better use of them. Thus they form a sort of dead-weight on the country, which can ill afford to be so handicapped in the race for improvement. Of course, individual exceptions exist, and deserve commendation, but the above is the rule." Lack of concern for inquiry by a British consul leaves the modern research question of why these Brazilians abstained from improvement. Wherein lay the logic of their operations? The quoted material appears in Palgrave's report on Uruguay for 1885–86, H. of C. 1887, LXXXVI, p. 871.

40. Hennessy, *Frontier in Latin American History*, p. 86.

41. Since the province of Rio Grande do Sul was embroiled in four major wars during the Empire (the Cisplatine campaign of 1817–28, the War of the Farrapos during the decade 1835–45, the campaigns in the Río de la Plata of 1849–52, and the Paraguayan War of 1864–70) and there were tensions along the borders for even longer, it is easy to see why the historiography of the region has been constructed around war and how the economic possibilities at times of peace have been underresearched. In addition, the impacts of the various armed struggles on the region's social and economic geographies remain fertile areas for further study.

42. Bradford Burns uses the "quixotic challenge" of the caudillo Aparicio Saravia to the power of central authority in Uruguay during 1903–4 as an ex-

of older, patriarchal values in the face of those who sought
y: "On horseback, wearing a white poncho and sombrero,
ne gaucho legacy. The struggle between Batlle and Saravia
conflict between those who favored a Europeanized future
ose who sought to draw on a Uruguayan-American past as a
:e. Batlle triumphed" (*Poverty of Progress*, p. 80). By the
tury, Uruguay had a modern and professionally equipped
ombat rural discord, but this had not been the case for most
:ntury. During the revolution of Venancio Flores against the
1 1863 (a revolution given critical aid by the ranchers of Rio
.e French minister to Uruguay made a fine two-camp analysis
:acterizing the struggle as "a battle between men without
s of the government) and horses without men (the Floristas)."
s been drawn from Barrán, *Apogeo y crisis del Uruguay*, p.

43. See the compelling study by Chasteen, *Heroes on Horseback*.

44. Leff, *Underdevelopment and Development in Brazil*.

45. Topik, *Political Economy of the Brazilian State*.

46. On the need for more economic case studies of Brazil, see especially Luz, "Brasil," p. 178.

47. Dauril Alden has drawn our attention to the absence of systematic investigation of the cattle economy of late colonial Rio Grande do Sul ("Late Colonial Brazil, 1750–1808," p. 879). In similar fashion, in the bibliographical article accompanying an essay on imperial Brazil, Richard Graham has argued that "particular sectors of the economy have not been studied in sufficient detail" and that "in contrast to the many studies on coffee and sugar, there are relatively few on the production of other crops or on the cattle economy" ("Brazil to the Paraguayan War," pp. 909–10). The otherwise commendably broad series of topics covered in Duncan and Rutledge, eds., *Land and Labour in Latin America*, includes only the minimum consideration of ranching. In their very perceptive article on the relationships between patterns of politics and violence on Latin America's ranching frontiers, Duncan Baretta and Markoff have argued that the social history of the ranch is the "most neglected variant" within the history of the great estate in Latin America ("Civilization and Barbarism," p. 618). For an important and recent regional study of Brazilian ranching, see Wilcox, "Cattle and the Environment in the Pantanal."

48. Love, "History—Pôrto Alegre," p. 91.

49. Scobie, *Argentina*; *Buenos Aires*; *Revolution on the Pampas*.

50. Within southern South America, the historiography of ranching is most developed in Argentina, but even for that country Slatta has argued the need for closer treatment of such basic topics as the evolution of land tenure systems and the formation of ranching elites (*Gauchos and the Vanishing Frontier*, p. 3); see also Jonathan C. Brown, *Socioeconomic History of Argentina*. For Rio Grande do Sul, the study of social and economic topics has hardly advanced enough yet to engender debate, let alone argument about the contours of the discussion, such as Slatta and Brown have developed for the social history of the gaucho in

nineteenth-century Argentina. On the urgent need for more systematic studies of the ranching areas north of Buenos Aires, see Garavaglia and Gelman, "Rural History of the Río de la Plata," p. 96.

51. The following studies have been especially important: Lobb, "Historical Geography of the Cattle Regions"; Pebayle, *Les Gauchos du Brésil*, especially pp. 129–82; Pesavento, *República velha gaúcha.*

52. The vast temporal sweep adopted in the studies by Lobb and Pebayle has the inevitable result that they provide very compressed treatments of the arrival of ranching technology. Lobb, in particular, explains centuries of development in the Campanha—from the Jesuit missions to the modern feedlot—through the idea of "minimum effort," a conclusion which calls for further examination of Rio Grande do Sul's ranching under specific sets of historical circumstances ("Historical Geography of the Cattle Regions," p. 209). The position of Rio Grande do Sul within national politics has been analyzed in detail for the Old Republic by Joseph Love. While this excellent book has been very widely quoted by scholars seeking information on the nature of the state's rural economy, it is much stronger on politics than on other aspects of the region. In this regard, the late James Scobie argued the following: "Although Love makes clear his intention not to probe into economic or social issues, in a study of political interaction it becomes difficult not to pay heed to these factors" (Scobie, Review of *Rio Grande do Sul*). For similar criticism, see Luz, "Brasil," p. 178.

53. Slatta, *Cowboys of the Americas.*

54. See ibid., pp. 16–17.

55. Clarence Jones published articles during the 1920's that were slight in a historical-cultural sense but that did make these broad divisions between a "core" and a "periphery" in the South American ranching zone. See, for example, his discussion of the Paraná-Uruguay grazing region, which included Rio Grande do Sul ("Agricultural Regions of South America").

56. Historical writing has used cross-sections for a very long time; in historical geography, these received a major boost as a working methodology in the 1930's through the work of H. C. Darby (*Historical Geography of England Before A.D. 1800*). On the combination of cross-sectional descriptions and chapters tracing social and economic changes, see also Broek, *The Santa Clara Valley, California*, and Darby, "Historical Geography," especially p. 140.

Chapter 1

1. Riograndino da Costa e Silva, *Notas à margem da história do Rio Grande do Sul*, pp. 79–110.

2. Müller, *Brasilien*, pp. 16–17, 124–25.

3. See Kohlhepp, "Strukturwandlungen in Nord-Paraná" p. 51, and "Donauschwaben in Brasilien," pp. 360, 376, 380.

4. Crosby, *Ecological Imperialism*, p. 149.

5. Pebayle, *Les Gauchos du Brésil*, p. 155.

6. Love, *Rio Grande do Sul and Brazilian Regionalism*, p. 130.

7. See the perceptive article by Pfeifer, "Kontraste in Rio Grande do Sul." This is reprinted with a preface by Gerd Kohlhepp in the collection of Pfeifer's essays, *Beiträge zur Kulturgeographie der Neuen Welt.*

8. See, for example, Chebataroff, "Regiones naturales del Uruguay y de Rio Grande del Sur."

9. On this theme, see especially the Brazilian translation of the classic geographical work by Roche, *Colonização alemã e o Rio Grande do Sul*, published originally as *La colonisation allemande et le Rio Grande do Sul*. See also Oberacker, "Colonização baseada no regime da pequena propriedade agrícola."

10. For a useful brief discussion of the senses of the term "Campanha," see Franco, "A Campanha," pp. 65–66.

11. Chebataroff, "Regiones naturales del Uruguay y de Rio Grande del Sur," p. 9.

12. Beek and Bramao, "Nature and Geography of South American Soils," pp. 88–89.

13. See Scobie, *Revolution on the Pampas*, pp. 15–18.

14. Beek and Bramao, "Nature and Geography of South American Soils," p. 110; Pfeifer, "Kontraste in Rio Grande do Sul," p. 292.

15. Knowledge of land-use potential has increased greatly with the survey of Rio Grande do Sul's natural resources conducted by the project RADAMBRASIL. See Fundação Instituto Brasileiro de Geografia e Estatística, *Levantamento de recursos naturais*.

16. Clark, "Impact of Exotic Invasion on the Remaining New World Mid-latitude Grasslands," p. 741; Sternberg, "Man and Environmental Change in South America," p. 427. Crosby has correctly observed that evidence on the floral changes of southern South America is "anecdotal, spotty, far from scientific" (*Ecological Imperialism*, p. 160).

17. Auguste de Saint-Hilaire (1779–1853) was a famous French botanist who undertook extensive scientific journeys in southern and central Brazil during the period 1816–22. A useful feature of his style is that he returned to certain themes constantly; the subtle subregional shifts in the quality of the vegetation was one of these (*Viagem ao Rio Grande do Sul*, especially pp. 113, 117, 123, 135–36, 181).

18. Roseveare, *Grasslands of Latin America*, pp. 53–55; see also Soriano et al., "Río de la Plata Grasslands," especially pp. 367–87.

19. Darwin, *Voyage of the Beagle*, p. 119.

20. See Pfeifer, "Kontraste in Rio Grande do Sul," pp. 291–93.

21. Danton Seixas, "Os nossos campos," *A Estância* (Oct. 1913): 228–29.

22. Roseveare, *Grasslands of Latin America*, p. 54.

23. Conde d'Eu, *Viagem militar ao Rio Grande do Sul*, pp. 45–46.

24. This point was stressed by C. A. M. Lindman, a Swedish botanist who undertook important research on the flora of Rio Grande do Sul (*Vegetação no Rio Grande do Sul*, pp. 4–5, 8).

25. Cesar, *História do Rio Grande do Sul*, p. 275.

26. There is a considerable specialized literature on the Luso-Spanish territorial rivalry. See especially Alden, *Royal Government in Colonial Brazil*, p. 71; Boxer, *Golden Age of Brazil*, pp. 239–54. For a very concise summary of the territorial rivalry over southern Brazil during the late colonial period, see Mansuy-Diniz Silva, "Portugal and Brazil," pp. 472–74.

27. As John Street expressed this, in his biographical study of the Uruguayan

caudillo Artigas, "Orientals and Portuguese or Brazilians were like Spaniards and Portuguese in Europe, or like oil and water: they would not mix" (*Artigas and the Emancipation of Uruguay*, p. 332).

28. Hemming, *Red Gold*, especially pp. 260–63.

29. Ibid., p. 463.

30. The vacarias contained large herds of feral cattle (*gado alçado*) that had multiplied on the open range. Apart from their famous mission settlements, the Jesuits also established an intricate complex of cattle stations, or primitive estâncias, east of the Uruguay River. "The sites chosen for the inauguration of breeding were the confluences of the rivers, the so-called *rincões*. There the herds were more or less detained by the waters and there they encountered the tenderest pastures." Animals that fell beyond Jesuit management (for whatever reasons) formed the bases of the herds in the vacarias. Santos, *Economia e sociedade do Rio Grande do Sul*, pp. 25–26, 61. The quotation above appeared originally in Cidade, "Rio Grande do Sul," p. 722.

31. The Duke of Cadaval had remarked in 1713 on the colony of Sacramento that when the Portuguese Crown held "dominions which yield gold counted in pounds and hundredweights, [it] should not bother about a few hides, which is all that colony produces." Boxer has noted that others shared this opinion, although not the Italian-born Jesuit André João Antonil (1649–1716), a famously astute observer of Brazil's economic structures (*Golden Age of Brazil*, pp. 245, 369–70).

32. These export figures for hides appear in Cesar, *História do Rio Grande do Sul*, p. 87.

33. Quoted in Boxer, *Golden Age of Brazil*, p. 241. "Jacum" derives from Tupi Indian and describes game-birds found in the woods (from the order *Galiformes*); "motuca" are large biting flies from the *Diptera* order (insects with one pair of wings); and "fascines" are bundles of tied sticks, used especially for fuel, ditch lining, and military engineering. André Ribeiro Coutinho's report can also be found in Rodrigues, *Continente do Rio Grande*, p. 32.

34. This priority was quite explicit, for example, in the instructions the captain-general of Rio Grande received from the colonial secretary during 1774, which ordered him to concentrate the defense on the town of that name: "His Majesty prizes much more the loss of a single league of territory in the southern part of Portuguese America than fifty leagues of exposed sertão in its interior" (quoted in Alden, *Royal Government in Colonial Brazil*, p. 468).

35. Hemming, *Red Gold*, pp. 462–86.

36. Félix de Azara (1746–1821) was an important figure in the Spanish Enlightenment. As a servant of the Spanish Crown in the Río de la Plata, he authored many perceptive works on the geography and natural history of southern South America. On Portuguese expansionism, see his "Memória rural do Rio da Prata," p. 56. This is a reprint of the translated version of Azara's manuscript published by Cortesão, *Do Tratado de Madri à conquista dos Sete Povos*, pp. 443–57. The original manuscript was signed on 9 May 1801 at Batovi (within the modern Riograndense county of São Gabriel), a settlement founded by Azara during his work determining the limits established through the Treaty of San Ildefonso.

37. The conquest of Missões actually took place after the cessation of hostilities in Iberia on 6 June 1801, but knowledge of the fact that the short-lived war had broken out did not reach Rio Grande until over a week after it had ended. Franco, *Origens de Jaguarão*, p. 17.

38. See Wiederspahn, *Bento Gonçalves e as guerras de Artigas*, p. 117.

39. Safford, "Politics, Ideology and Society in Post-Independence Spanish America," pp. 411–12.

40. The interests of the Portuguese court in the Banda Oriental and the nature of Brazil's territorial controversy with the United Provinces of the Río de la Plata have been summarized by Seckinger, *Brazilian Monarchy and the South American Republics*, pp. 59–73, 144–51. For an introduction to Brazil's foreign-policy issues in the Plata from the creation of Uruguay to around mid-century, see Sousa, "O Brasil e o Rio da Prata de 1828 à queda de Rosas."

41. See the report on the Banda Oriental made by Thomas Samuel Hood to George Canning, Montevideo, 16 Aug. 1824, printed in Humphreys, ed., *British Consular Reports*, especially pp. 75, 77–78, 81, 86.

42. Lavalleja, a former follower of Artigas, was one of a series of political leaders who sprang from the rural zones of the Banda Oriental. See Halperín-Donghi, *Politics, Economics and Society in Argentina*, pp. 270–71.

43. See Leitman, "Socio-economic Roots of the Ragamuffin War," p. 107.

44. Ibid., p. 158.

45. In Rio Grande do Sul itself, the dominant trend in the historiography of the war during recent decades has been to view the Farrapos as having republican and federalist but not separatist leanings. Walter Spalding, a leading proponent of this approach, viewed the origins of the Farroupilha Revolt as lying in the end of the colonial period, when the Crown was paying very little attention to the economic well-being of the southern frontier (*Revolução Farroupilha*, pp. 12–13).

46. The decrees through which the Portuguese Crown attempted to control the dimensions of concessions of land to individuals are reviewed in Santos, *Economia e sociedade do Rio Grande do Sul*, pp. 44–45.

47. The Azorian presence in Rio Grande do Sul has been used to explain the cultural differences that separated the gaúchos of that region from their Spanish-speaking neighbors. The argument is based on the theory that the Azorians imparted conservative and sedentary customs to the inhabitants of Brazil's southern frontier. Ibid., p. 39.

48. Lobb, "*Sesmaria* in Rio Grande do Sul," p. 62.

49. See Pebayle, *Les Gauchos du Brésil*, p. 136.

50. This is based on data presented in Santos, *Economia e sociedade do Rio Grande do Sul*, p. 56.

51. Lobb, "*Sesmaria* in Rio Grande do Sul," p. 49.

52. See, for example, the copy of a sesmaria grant contained within the inventory of deceased: Antonio Martins da Cruz Jobim, Barão de Cambaí; executrix: Ana de Sousa Brasil, Baronesa de Cambaí; São Gabriel, 1869, N 210, M 11, E 107, APRGS. The recipient of the sesmaria described within this inventory (which comprised some of the very best ranchland within Rio Grande do Sul) was José Martins de Oliveira, who described himself as an "inhabitant of

the Rio Pardo frontier." The viceroy, the Conde de Rezende, conferred the lands described in the grant on 18 Oct. 1798 in Rio de Janeiro. The recipient had asked that the viceroy concede lands to him through legitimate title, based on the argument that he had "settled more than ten years ago with the authorization of the military commanders of the same frontier on some lands where he lives established." In his book on the early settlement in the Jaguarão frontier region, Sérgio da Costa Franco has argued against reading too much into sesmaria grants. When lands with an area of less than a square league were conceded along the frontier, a sesmaria letter was not required. Such was the flexibility along the frontier that he has even documented a case of a Brazilian who managed to settle within territory controlled by the Spanish with the complicity of their authorities (*Origens de Jaguarão*, pp. 13, 16).

53. In this context, "campanha" is used in the sense of interior. Roscio, "Compêndio noticioso," p. 132. This is a reprint of the version of this manuscript transcribed by General João Borges Fortes and published in *RIHGRGS* 87 (1942): 29–56.

54. Magalhães, "Almanack da vila de Porto Alegre," p. 84. On the same topic, see also Gonçalves Chaves, *Memórias ecônomo-políticas*, especially pp. 181–91.

55. Kerst, "Die brasilische Provinz Rio Grande do Sul," p. 104. Kerst served in Rio Grande do Sul as a captain in the Brazilian engineering corps from 1826 to 1831. His detailed chorography of the province is important and worthy of a modern critical edition.

56. At least as far as the Jaguarão frontier is concerned, not all of the petitioners for land spent much time on it. Frequent instances appear of people located in the regional urban centers who held land on the frontiers. For example, the father of the future Farrapo leader Bento Gonçalves da Silva moved into the Camaquã area prior to 1780. There he successively acquired several estâncias which he managed from Porto Alegre. Franco, *Origens de Jaguarão*, pp. 11, 13; Wiederspahn, *Bento Gonçalves e as guerras de Artigas*, p. 116.

57. See Cesar, *Contrabando no sul do Brasil.*

58. There were further serious droughts in northeast Brazil in 1791–93 and 1809–10. See Bauss, "Rio Grande do Sul in the Portuguese Empire," p. 522.

59. Machado, "A charqueada," p. 121.

60. Cesar, *História do Rio Grande do Sul*, p. 208.

61. Ibid.

62. Mawe, *Travels in the Interior of Brazil*, pp. 444–45.

63. Bauss, "Rio Grande do Sul in the Portuguese Empire," p. 535.

64. One Uruguayan government report estimated the number of animals taken out of that country to Brazil during the Cisplatine War at 24 million. See Leitman, "Socio-economic Roots of the Ragamuffin War," p. 95. This figure is almost certainly exaggerated.

65. Wiederspahn, *Bento Gonçalves e as guerras de Artigas*, p. 117.

Chapter 2

1. Manuel Luís Osório, the Marquês do Herval (1808–79), was a gaúcho and one of Brazil's most distinguished soldiers during the nineteenth century.

As a Liberal, he played a major role in national politics after the Paraguayan War. See Love, *Rio Grande do Sul and Brazilian Regionalism*, pp. 21–22. The conversation quoted appears in Fontoura, *Borges de Medeiros e seu tempo*, p. 376.

2. Luccock, *Notes on Rio de Janeiro*, pp. 167, 216.

3. Isabelle, *Voyage a Buénos-Ayres et a Porto-Alègre*, pp. 445–46. Examining the mechanics of landownership for Buenos Aires Province during the early nineteenth century, Jonathan C. Brown has concluded that "for the cattleman, sound economic reasons existed for settling the virgin prairies with large landholdings. The nature of raising cattle, with the traditional rural technology and meager labor pools of the era, demanded expansive pasturages" (*Socioeconomic History of Argentina*, p. 3).

4. V.S.L., art. 6. José Feliciano Fernandes Pinheiro, Visconde de São Leopoldo (1774–1847), was the first president of the province of Rio Grande do Sul (1824–26). Brazilian-born, he was educated at the University of Coimbra in Portugal. After graduation, he worked during 1799–1801 as a translator of studies from English and French (including works on agricultural improvement) in the literary establishment designed to educate Brazilian landowners that was directed by José Mariano da Conceição Veloso, a Brazilian-born friar. See Aurélio Porto's preface in Pinheiro, *Anais da Província de S. Pedro*, pp. vii-xxv; Galloway, "Agricultural Reform in Late Colonial Brazil," pp. 774–75.

5. João Antonio Pereira Martins, Visconde de Serro Azul (1767–1847), was the maternal grandfather of Gaspar Silveira Martins, the famous politician of the late Empire, who was born in northern Uruguay in 1834. Almost all of the Visconde de Serro Azul's ranches are reckoned to have been bought; he did not receive them as sesmaria grants. They ran from the Bagé region into the heart of Uruguay. The Visconde de Serro Azul reportedly said: "If God gives me a long enough life and with his help, I shall go from 'Candiota' to Montevideo within my own lands." And on another occasion: "Houses as many as suffice, land to the limits of the horizon." It comes as no surprise to learn that he lent major material support to the conduct of the Cisplatine campaign. See Carvalho, *Nobiliário sul-rio-grandense*, pp. 262, 265, 277. Even landed wealth on this scale did not compare with that accumulated by the leading coffee factors and growers of Rio de Janeiro. See, for example, Richard Graham's biographical sketch of Antonio Clemente Pinto, Barão de Nova Friburgo ("Brazil to the Paraguayan War," p. 765).

6. Lynch, *Argentine Dictator*, pp. 70–72; Jonathan C. Brown, "Nineteenth-Century Argentine Cattle Empire."

7. See Chasteen, *Heroes on Horseback*, pp. 50–52.

8. See Correa da Câmara, "Ensaios statisticos do Rio Grande do Sul," pp. 22–23.

9. Chasteen rejects variations in land quality and land use as a potential explanation ("Background to Civil War," p. 745). However, the eastern parishes in his sample are both in the Serra do Sudeste, a very different physical environment from the Campanha proper. Even today, average property sizes in the Campanha are around 3.5 times larger than in the Serra do Sudeste. See the

data in Benetti, "Agropecuária na região sul do Rio Grande do Sul," p. 185, table 3, and map 1, following p. 227.

10. The landholding history of the Simões Lopes, a family with charqueada interests, makes an excellent case in point. Portuguese-born Commander João Simões Lopes established the Estância da Graça in Pelotas around 1800. His son, the Visconde da Graça, who died in 1893, left ranches in São Gabriel, Uruguaiana, and Durazno, Uruguay, in addition to the family business-seat on the coast. See Reverbel, *Capitão da Guarda Nacional*, especially pp. 15, 28–33.

11. Alden, "Late Colonial Brazil, 1750–1808," p. 642; Leitman, "Socio-economic Roots of the Ragamuffin War," p. 93. Charque followed wheat and hides in the region's exports in 1814 but headed the list by 1821.

12. Roscio, "Compêndio noticioso," p. 135. One of the factors used to explain the evolution of cattle ranching within medieval Iberia was its relative flexibility in regions of prolonged political instability. See Bishko, "Peninsular Background of Latin American Cattle Ranching," p. 497.

13. Dante de Laytano argues that the Azorian settlers had usually rejected agriculture by the second generation, transforming themselves into *"fazendeiros and cattle-breeders, estância owners"* ("Colonização açoriana no Rio Grande do Sul," p. 392. See also Saint-Hilaire, *Viagem ao Rio Grande do Sul*, pp. 110, 120. There are discussions of the decline of wheat in Pebayle, *Les Gauchos du Brésil*, pp. 150–51, and Lobb, "Historical Geography of the Cattle Regions," pp. 148–49.

14. The nature of North American competition is revealed in Gregory Brown's recent study "Impact of American Flour Imports on Brazilian Wheat Production," especially pp. 319, 326, 329–30, 333–34.

15. See Luccock, *Notes on Rio de Janeiro*, pp. 195–96; Saint-Hilaire, *Viagem ao Rio Grande do Sul*, pp. 29, 80–81, 173, 179.

16. "The indolence of the inhabitants of these places (with very honorable exceptions) has become proverbial and makes a contrast with the benignity of the climate, the beauty and fecundity of the lands." CM of Alegrete, doc. 729, 28 June 1852, AHRGS.

17. CM of Alegrete, doc. 1038, 11 Dec. 1862, AHRGS.

18. CM of Alegrete, doc. 1097, 30 Nov. 1868, AHRGS.

19. CM of São Gabriel, doc. 30, 7 Jan. 1870, AHRGS.

20. Petition witnessed by Clementino Machado dos Santos, public notary, Dom Pedrito, 12 Mar. 1867; CM of Bagé, doc. 390, AHRGS.

21. "Representação dos moradores da Freguesia do Estreito a S.M.I., queixando-se das condições exigidas pelo arrendatário da Fazenda Bojuru para a utilização das terras, e solicitando a edificação de uma capela na região," Rio Grande, 4 Sept. 1823; manuscripts, II-35, 36, 1 n. 3, BN.

22. CM of Alegrete, doc. 729, 28 June 1852, and CM of Bagé, doc. 18, 16 Mar. 1847, AHRGS.

23. V.S.L., arts. 2, 7. See also Dreys, *Notícia descritiva da província do Rio Grande*, p. 131. Dreys was a Frenchman who lived in Rio Grande do Sul for at least the period 1818–28.

24. CM of São Gabriel, doc. 312a, 1859, AHRGS. To establish the sheer scale of the cattle holdings in this single county, it helps to note that the whole

neighboring region of Corrientes recorded 673,390 cattle in a census of 1854; see Whigham, "Cattle Raising in the Argentine Northeast," p. 331.

25. CM of São Gabriel, doc. 30, 7 Jan. 1870, AHRGS. For a brief review of the role of sheep in the traditional ranching economy of Rio Grande do Sul, see Bell, "Aimé Bonpland and Merinomania," pp. 312–14.

26. Inventory of deceased: Antonio Guterres Alexandrino; executrix: Ana Joaquina Flora; Alegrete, 1853, N 117, M 8, E 65, APRGS.

27. Oliveira Belo, "Diário de uma excursão eleitoral," p. 26. The journey described was undertaken in 1856.

28. Roscio, "Compêndio noticioso," p. 135.

29. Winter cattle loss was a serious problem. A rancher in the western interior related that since the drought of 1840 cattle had died in the hundreds even on small estâncias. He blamed the mortality on overstocking and "the erroneous system of not a few ranchers" of thinking only of augmenting their herds without giving any thought to the pace at which the grasses grew. The grasslands of Uruguay were still the dumping ground ("lugar de descarga") for the "very damaging superabundance" of cattle in Rio Grande do Sul. See Francisco de Sá Brito to Joaquim Antão Fernandes Leão, Rincão de Paipasso, 10 Dec. 1859; correspondence of the presidents, maço 30, AHRGS.

30. In the well-stocked landscape of the missions, the Jesuit Anton Sepp claimed in 1698 that the grass was around 66 centimeters (a *côvado*) high. Quoted in Santos, *Economia e sociedade do Rio Grande do Sul*, p. 60. Based on the comments of a Lübeck doctor who crossed the Campanha in 1858, the grasses of western Rio Grande do Sul were still sometimes taller than horse and rider. Avé-Lallemant, *Viagem pela província do Rio Grande do Sul*, p. 203.

31. "The pastures consist of sward that arises and grows spontaneously; on new ranges it grows up with such opulence that it comes to cover and conceal the animal. However, it diminishes and is crushed down in proportion with the number of animals that it sustains and when in excess, they waste away and die of hunger. The ranchers know this danger from practice and from it comes the adage— *the range already cannot cope with the weight of the cattle* = *you have to acquire new lands to move them to*" (emphasis in original). V.S.L., art. 2.

32. Ibid.

33. See Dreys, *Notícia descritiva da província do Rio Grande*, p. 132; Saint-Hilaire, *Viagem ao Rio Grande do Sul*, p. 162. Dreys claimed it was unnecessary to feed salt to cattle in Rio Grande do Sul, but his comments clearly refer to the Litoral and not the interior.

34. V.S.L., art. 16.

35. The set of *posturas* (county regulations) drawn up by the councillors of Alegrete in 1849 informed ranchers that it was no longer acceptable to use the sign "commonly known as *tronxo*," which consisted of amputating both of an animal's ears. See section 108 of "Costeio das Fazendas e registo das tropas"; CM of Alegrete, doc. 556, 9 June 1849, AHRGS.

36. Dreys, *Notícia descritiva da província do Rio Grande*, p. 130.

37. Kerst, "Die brasilische Provinz Rio Grande do Sul," p. 309.

38. Cesar, *Conde de Piratini*, p. 42.

39. Baguet, *Rio Grande do Sul et le Paraguay*, p. 66. This book was based

on a journey made during 1845–46. Baguet was a Belgian who lived in Brazil between 1842 and 1874. During part of that time, he served as tutor to the imperial princesses. On returning to Belgium, he was appointed honorary Brazilian consul in Antwerp.

40. Roscio, "Compêndio noticioso," p. 135.

41. V.S.L., arts. 4, 5.

42. See Nunes and Cardoso Nunes, *Dicionário de regionalismos do Rio Grande do Sul*, pp. 515–17.

43. See Roscio, "Compêndio noticioso," p. 135.

44. V.S.L., art. 3.

45. Luccock, *Notes on Rio de Janeiro*, p. 217.

46. Magalhães, "Almanack da vila de Porto Alegre," p. 79.

47. Cesar, *Conde de Piratini*, pp. 38–39.

48. V.S.L., art. 3. Luccock claimed in the early part of the century that the rodeios were taking place in northern Uruguay once a year but in the Litoral of Rio Grande do Sul "as circumstances may require" (*Notes on Rio de Janeiro*, pp. 167, 216).

49. See the official letter from Manoel Pereira da Silva Ubatuba to president João Sertorio, as published in *A Reforma*, 22 July 1869.

50. See, for example, section 111 of Alegrete's municipal code; CM of Alegrete, doc. 556, 9 June 1849, AHRGS.

51. Slatta, *Gauchos and the Vanishing Frontier*, pp. 8, 87; Jonathan C. Brown, *Socioeconomic History of Argentina*, pp. 172–73.

52. See Nunes and Cardoso Nunes, *Dicionário de regionalismos do Rio Grande do Sul*, pp. 68–69, 258–59.

53. Samuel Gottfried Kerst observed estância boys training with miniature versions of the bolas on dogs, sheep, and chairs ("Die brasilische Provinz Rio Grande do Sul," p. 315).

54. Dreys, *Notícia descritiva da província do Rio Grande*, pp. 148–49, 161. For other descriptions of these tools, see Baguet, *Rio Grande do Sul et le Paraguay*, pp. 56–57; Luccock, *Notes on Rio de Janeiro*, p. 205.

55. Cesar, *Conde de Piratini*, p. 45.

56. Claudino Mariano de Sales, capataz, to Maria Alves de Oliveira, Alegrete, 16 Dec. 1870. This letter is contained within the inventory of deceased: Maria Eleuteria; executrix: Maria Alves de Oliveira with the help of her husband, Justino Torres Filho; Alegrete, 1870, N 304, M 24, E 65, APRGS.

57. Inventory of deceased: Antonio Martins da Cruz Jobim, Barão de Cambaí; executrix: Ana de Sousa Brasil, Baronesa de Cambaí; São Gabriel, 1869, N 210, M 11, E 107, APRGS.

58. V.S.L., art. 15.

59. For details of the presence of slaves on the ranches of Argentina, see Slatta, *Gauchos and the Vanishing Frontier*, pp. 34–35, 203; Jonathan C. Brown, *Socioeconomic History of Argentina*, pp. 42, 163, 187–88. For further studies, see also the discussion in Garavaglia and Gelman, "Rural History of the Río de la Plata," pp. 87, 92.

60. For a graphic description of the shortage and high cost of labor, see Luccock, *Notes on Rio de Janeiro*, p. 223. On the movement of slaves to Rio

Grande do Sul, see Alden, "Late Colonial Brazil, 1750–1808," p. 612; Karasch, *Slave Life in Rio de Janeiro*, pp. 51–52; Mawe, *Travels in the Interior of Brazil*, p. 445.

61. Cesar, *História do Rio Grande do Sul*, pp. 31, 235.

62. See, for example, Franco, "Esquema sociológico da fronteira," p. 47. The clearest introductory discussion of the various roles of slaves within Rio Grande do Sul's agro-pastoral economy can be found in Cardoso, *Capitalismo e escravidão no Brasil meridional*, pp. 60–69.

63. In the early part of the nineteenth century, some authors argued that there were important differences between the labor sources on the ranches of Rio Grande do Sul and the Plata. John Mawe, for example, claimed that "the Spaniards have Peons on their farms, who are more nearly allied to the Indians than to them, whereas the Portuguese have Creolians, bred up to the business, or expert negroes, who are inferior to none in this labour" (*Travels in the Interior of Brazil*, p. 448). Nicolau Dreys maintained that while the peons were sometimes black slaves, the use of Indians or paid labor ("gaúchos assalariados") was more frequent (*Notícia descritiva da província do Rio Grande*, p. 130).

64. Leitman hoped that his discussion of the evidence for slave cowboys would serve "as but the beginning of . . . a series of new dialogues on slavery in southern Brazil" ("Slave Cowboys in the Cattle Lands," p. 176). For a revisionist view of the supposed egalitarian attitudes adopted by the Farrapos toward the slaves in the rebel army, see also his article "Black Ragamuffins." Some years after the Farroupilha, a representative from Rio Grande do Sul in the chamber of deputies noted that the ranchers encountered difficulty in working the estâncias during the revolt, when slaves were quick to take up arms. This led them to use free peons instead, who used the shortage of available labor to bid up their rates. Speech by Jacintho de Mendonça, session of 1 July 1853, as reported in O *Mercantil*, 3 Aug. 1853.

65. Brazil, *Relatorio da repartição dos negocios estrangeiros* (1851), p. 41.

66. The statistical list does not bear a date, but it is clear from the manner in which the counties are listed that the survey was conducted around 1860. Even Leitman, probably being cautious not to overstate a "highly speculative" case, emphasized that "mestizos and Indian gauchos and peons were always numerically and economically more important than the slave cowboy" ("Slave Cowboys in the Cattle Lands," p. 171). For further evidence that this was not the case, see Franco, *Origens de Jaguarão*, p. 95.

67. Maestri Filho, *Escravo no Rio Grande do Sul*, p. 53.

68. Brazil, *Relatorio da repartição dos negocios estrangeiros* (1851), p. 41.

69. Inventory of deceased: Benigno José de Souza; executrix: Candida Olinda de Freitas (widow); Bagé, 1849, N 62, M 3, E 38, APRGS.

70. Cesar, *Conde de Piratini*, p. 17; Freitas, O *capitalismo pastoril*, pp. 33–36.

71. Freitas, O *capitalismo pastoril*, p. 35.

72. José Antonio Pimenta Bueno to Paulino José Soares de Souza, 5 Aug. 1850; RGS, *avisos recebidos*, 1850–52, 310, 1, 1, AHI. See also Chasteen, *Heroes on Horseback*, p. 96.

73. Inventory of deceased: João Silveira Gularte; executor: Constantino José de Freitas; Bagé, 1849, N 58, M 3, E 38, APRGS.

74. Brazil, *Relatorio da repartição dos negocios estrangeiros* (1851), p. 41.

75. Saint-Hilaire could not understand why the ranchers of Missões used paid labor instead of African slaves. He was told that the Indian women preferred black men as lovers to any others; in turn, these Indian women infected the blacks with venereal disease, which killed them (*Viagem ao Rio Grande do Sul*, p. 187).

76. For a brief account of the regional contrasts around 1860, see Franco, *Júlio de Castilhos e sua época*, pp. 1–3.

77. J. J. Martinez, "Observaciones industriales," *RARU*, no. 78 (1 Mar. 1876): 27.

78. Inventory of deceased: José Maria da Gama Lobo Coelho d'Eça, Barão de Saican; executrix: Baronesa de Saican; São Gabriel, 1873, N 9, M 1, E 108, APRGS. The Law of Free Birth (or Rio Branco Law) of September 1871 freed (with some reservations) the children born of slave mothers after that date. The law was a major step in Brazil's gradual abolition of slavery.

79. Inventory of deceased: Francisco de Assis Brazil; executrix: Joaquina Theodora Brazil; São Gabriel, 1872, N 247, M 12, E 107, APRGS.

80. V.S.L., art. 3. Luccock claimed that "to each three square leagues are allotted four or five thousand head of cattle, six men and a hundred horses" (*Notes on Rio de Janeiro*, p. 216).

81. V.S.L., art. 3.

82. Brito, *Trabalhos e costumes dos gaúchos*, p. 57.

83. Inventory of Benigno José de Souza, Bagé, 1849, APRGS.

84. V.S.L., art. 5.

85. Saint-Hilaire, *Viagem ao Rio Grande do Sul*, pp. 47, 90, 117.

86. Inventory of Antonio Guterres Alexandrino, Alegrete, 1853, APRGS.

87. Province of RGS, *Relatório, 2-10-1854*, p. 45. (An indispensable source for regional studies of the Brazilian Empire, presidential reports to the provincial legislative assemblies, while printed, are not widely available today. For Rio Grande do Sul, there is a good collection in the AHRGS.) Cattle disease affected a much broader area than Rio Grande do Sul; see Whigham, "Cattle Raising in the Argentine Northeast," pp. 328–29.

88. Domingos José de Almeida to Joaquim Antão Fernandes Leão, Pelotas, 20 Nov. 1859, and Joaquim Gonçalves da Silva to Fernandes Leão, S. João Batista de Camaquam, 20 Dec. 1859; correspondence of the presidents, maço 30, AHRGS. Domingos José de Almeida (1797–1871) was one of the leading entrepreneurs of Pelotas. A Farrapo, he served the Riograndense Republic as its minister of finance, among other roles. Feliciano Antonio de Moraes to Fernandes Leão, Bagé, 7 Jan. 1860; correspondence of the presidents, maço 31, AHRGS.

89. "There are (and one ought to say it with surprise) ranchers who sell [diseased cattle], merchants who buy them, and charqueadores who slaughter them; all of these cry that the business is lost without noticing that they themselves are the principal cause of this ill"; Province of RGS, *Relatório, 1º.-06-1849*, p. 10. Francisco José de Sousa Soares de Andréa, Barão de Caçapava

(1781–1858), came to Brazil with the removal of the Portuguese court in 1808. He had an important career as a professional administrator.

90. V.S.L., art. 5.

91. CM of Bagé, doc. 22, 5 May 1847, and CM of Alegrete, doc. 495, 10 July 1847, AHRGS.

92. See CM of Bagé, docs. 20, 20a, 29 Mar. 1847, AHRGS.

93. Province of RGS, *Relatório*, 2-10-1854, p. 45.

94. Province of RGS, *Relatório*, 1°.-10-1852, p. 33. During his journey through Rio Grande do Sul in 1858, Avé-Lallemant offered the view that although the herds were "more or less restored," they had not regained the state achieved before 1835. Like another European traveler some years later, the Conde d'Eu, Avé-Lallemant was particularly unimpressed with the condition of Rio Grande do Sul's horses (*Viagem pela província do Rio Grande do Sul*, pp. 363, 365–66).

95. Given the abundance and low value of cattle earlier in the century, robbery was also part of the traditional way of life. Around 1830, Kerst maintained that even landowners fed their dependents on their neighbors' cattle ("Die brasilische Provinz Rio Grande do Sul," pp. 318–19).

96. See the news from the frontier contained in a letter from Pelotas, dated 27 Nov. 1848, published in the *Diário do Rio Grande*, 29 Nov. 1848.

97. José Antônio Pimenta Bueno, president of RGS, to the Minister of Foreign Affairs, Porto Alegre, 30 Sept. 1850; RGS, avisos recebidos, 1850–52, 310, 1, 1, AHI.

98. Letter from Alegrete dated 12 Aug. 1853, published in O *Mercantil*, 15 Sept. 1853. On the same theme, see also the letter from Patricio Vieira Rodrigues to the Barão de Muritiba, P. of P., Freguesia das Dores, 22 Feb. 1856; correspondence of the presidents, maço 27, AHRGS.

99. "Complaints against cattle-theft are repeated from all points of the province. They are becoming so frequent and on such a large scale that one could say private authority (*domínio particular*) is a fanciful conception (*chimera*) today. Only the blind public could seriously take responsibility for it." CM of Bagé, doc. 234, 14 Jan. 1857, AHRGS.

100. CM of Alegrete, doc. 936, 10 Jan. 1858, AHRGS.

101. Vicuña Mackenna, *La Argentina en el año 1855*, p. 133, quoted in Scobie, *Revolution on the Pampas*, p. 9.

102. Jonathan C. Brown, "Nineteenth-Century Argentine Cattle Empire," pp. 176–77.

103. Saint-Hilaire, *Viagem ao Rio Grande do Sul*, pp. 123–24.

104. This distinction is lost in some modern historical writing. While making a successful case for a more historically informed analysis of labor in cattle ranching economies, Ricardo Salvatore confuses understanding of Rio Grande do Sul by treating ranchers and charqueadores as a single group ("Modes of Labor Control in Cattle-Ranching Economies," pp. 447–48, 451).

105. Correa da Câmara, "Ensaios statisticos do Rio Grande do Sul," pp. 22–23; Pedro de Alcântara Bellegarde, "Apontamento sobre a República do Paraguai e sobre a Província do Rio Grande do Sul," Asunción, 3 June 1849; manuscripts, 7, 4, 59, BN.

106. Correa da Câmara, "Ensaios statisticos do Rio Grande do Sul," pp. 22–23. Antônio Manoel Correa da Câmara (1783–1848) served as a soldier and had a distinguished diplomatic career during the early nineteenth century. Between 1845 and his death, he was in charge of collecting statistics in Rio Grande do Sul.

107. Jonathan C. Brown, "Nineteenth-Century Argentine Cattle Empire," pp. 161, 178.

108. Cesar, *Conde de Piratini*, p. 12.

109. Irineu Evangelista de Sousa, Barão and later Visconde de Mauá (1813–89), was generally recognized as Brazil's leading industrial capitalist during the middle decades of the nineteenth century. He was born in Arroio Grande, Rio Grande do Sul. For a brief account of Mauá's career in such important areas as railway promotion and bank formation, see Richard Graham, *Britain and the Onset of Modernization in Brazil*, pp. 187–96. An obituary of the Barão de Cambaí can be found in *A Reforma*, 8 Aug. 1869.

110. Inventory of the Barão de Cambaí, São Gabriel, 1869, APRGS.

111. João Theodoro de Mello Souza Barreto, Caixa Econômica de São Gabriel, to Angelo Moniz da Silva Ferraz, P. of P., São Gabriel, 18 Sept. 1858; correspondence of the presidents, maço 29, AHRGS.

112. See the speech made by deputy Teixeira de Almeida in the provincial assembly, session of 27 Dec. 1859, as reported in the *Correio do Sul*, 7 Feb. 1860.

113. Inventory of deceased: Manoel Joaquim do Couto and Potenciana Joaquina do Couto; executor: Agostinho Pereira de Carvalho; Alegrete, 1848, N 92, M 7, E 65, APRGS.

114. Frances, *Beyond the Argentine*, pp. 60–61; see also Conde d'Eu, *Viagem militar ao Rio Grande do Sul*, p. 46.

115. Avé-Lallemant, *Viagem pela província do Rio Grande do Sul*, pp. 329–30.

116. J. J. Martinez, "Observaciones industriales," *RARU*, no. 78 (1 Mar. 1876): 27.

117. The move toward keeping houses in the towns was already evident at the beginning of the century. In Rio Grande, Luccock had noted that "some of the houses, belonging to persons who reside on their estates, are seldom occupied except at religious festivals" (*Notes on Rio de Janeiro*, p. 174). On the trend of richer estancieiros wintering in the towns, see Kerst, "Die brasilische Provinz Rio Grande do Sul," p. 326.

118. Cesar, *Conde de Piratini*, pp. 22, 45.

119. Inventory of the Barão de Saican, São Gabriel, 1873, APRGS.

120. Inventory of João Silveira Gularte, Bagé, 1849, APRGS.

121. Baguet, "Rio Grande-do-Sul," p. 399.

122. On the "alimentary regime," see Dreys, *Notícia descritiva da província do Rio Grande*, pp. 169–70; Kerst, "Die brasilische Provinz Rio Grande do Sul," pp. 325–27; Cesar, *Conde de Piratini*, pp. 39–40.

123. CM of São Gabriel, doc. 312a, 1859, AHRGS.

124. Inventory of Antonio Guterres Alexandrino, Alegrete, 1853, APRGS.

125. Kerst, "Die brasilische Provinz Rio Grande do Sul," p. 325.

126. Avé-Lallemant, *Viagem pela província do Rio Grande do Sul*, p. 207.

127. CM of Alegrete, doc. 729, 28 June 1852, and doc. 1038, 11 Dec. 1862, AHRGS.

128. The accounts range from mid-April to early July 1870; inventory of the Barão de Cambaí, São Gabriel, 1869, APRGS.

129. Cesar, *Conde de Piratini*, pp. 43–44; inventory of the Barão de Cambaí, São Gabriel, 1869, APRGS.

130. See, for example, the descriptions in Luccock, *Notes on Rio de Janeiro*, p. 198; Avé-Lallemant, *Viagem pela província do Rio Grande do Sul*, pp. 192, 232, 246; Conde d'Eu, *Viagem militar ao Rio Grande do Sul*, pp. 31–32, 46.

131. See, for example, Conde d'Eu, *Viagem militar ao Rio Grande do Sul*, p. 47.

132. Baguet, "Rio Grande-do-Sul," p. 407.

133. Oliveira Belo, "Diário de uma excursão eleitoral," p. 44.

134. CM of Bagé, doc. 31, 16 Sept. 1847, AHRGS.

135. Oliveira Belo, "Diário de uma excursão eleitoral," p. 44.

136. Avé-Lallemant, *Viagem pela província do Rio Grande do Sul*, p. 235.

137. CM of Bagé, doc. 163, 26 Feb. 1853, AHRGS. For the argument that ambulant trade robbed the towns of any true dynamism, see CM of São Gabriel, "Relatório apresentado à Câmara Municipal da cidade de São Gabriel pelo Vereador Presidente, Geraldo de Faria Corrêa (manuscrito)," 1873, AHRGS.

138. Province of RGS, *Relatório, 5-11-1858*, p. 41. For an interesting analysis of contraband, see the confidential letter from John Morgan, British consul at Rio Grande do Sul to José Antônio Pimenta Bueno, 21 Mar. 1850; consular correspondence, maço 10, AHRGS.

139. Avé-Lallemant, *Viagem pela província do Rio Grande do Sul*, pp. 288, 308, 326–27; see also Conde d'Eu, *Viagem militar ao Rio Grande do Sul*, p. 63.

140. Thomas Whigham makes his summary judgment on the state of Riograndense ranching using data drawn from São Borja, one of the province's poorer ranching zones (*Politics of River Trade*, p. 171).

Chapter 3

1. Portions of this chapter draw from my brief study of the Riograndense slaughtering-plants in comparative perspective ("Early Industrialization in the South Atlantic").

2. Pedro de Alcântara Bellegarde, "Apontamento sobre a República do Paraguai e sobre a Província do Rio Grande do Sul," Asunción, 3 June 1849, manuscripts, 7, 4, 59, BN.

3. CM of Pelotas, doc. 425, 15 Mar. 1856, AHRGS. The exact number of charqueadas working at any specific time is difficult to determine. João Simões Lopes Neto claimed in 1911 that around forty different plants had been established in Pelotas at different times and a hundred in the entire province. Some two hundred different companies had used the forty Pelotas establishments. Even by 1911, when Pelotas celebrated its first centenary, many of the charqueadas had left few vestiges ("Noticia sobre a fundação das xarqueadas," pp. 45–46).

4. CM of Pelotas, doc. 378a, 21 Aug. 1854, AHRGS; Whigham, "Cattle Raising in the Argentine Northeast," p. 326.

5. CM of Pelotas, doc. 378a, 21 Aug. 1854, AHRGS.

6. "In this way, the fate of the business remained in exclusive dependence on the 'buyer's eye.' It was said then that if the 'owner's eye' fattens the ox in the invernada, the 'buyer's eye' responds through its yield in the charqueada. Appealing for the weigh scales (balança) was proof of being a *maturrango* [literally, "a poor horseman"; this word carries connotations of impracticality] or 'Bahian,' an undistinguished, even humiliating, condition in the pastoral environment of that time. The compulsory use of the weigh scales in transactions over cattle for slaughter was only introduced with the arrival of the gringos of the frigoríficos." Reverbel, *Capitão da Guarda Nacional*, pp. 19–20.

7. Luccock, *Notes on Rio de Janeiro*, p. 213.

8. Saint-Hilaire, *Viagem ao Rio Grande do Sul*, p. 73; Avé-Lallemant, *Viagem pela província do Rio Grande do Sul*, pp. 407–8.

9. Dreys, *Notícia descritiva da província do Rio Grande*, p. 134.

10. Saint-Hilaire, *Viagem ao Rio Grande do Sul*, pp. 75–76; V.S.L., art. 24.

11. 1 arroba = 15 kilos.

12. V.S.L., art. 12. Jonathan Brown has data showing that the average steer in Buenos Aires produced a hide of 27 kilograms around mid-century. Even bearing in mind that there were larger creole cattle in Buenos Aires than in Rio Grande do Sul, this weight seems too high. During the 1890's, hides taken from creole cattle in Uruguay ranged from 13.6 to 18.2 kilograms when sun-dried, weights that make the Visconde de São Leopoldo's data look plausible. Jonathan C. Brown, *Socioeconomic History of Argentina*, p. 111; Grenfell's report on Uruguay for 1889–91, H. of C. 1893–94, XCVIII, p. 205.

13. Vereker's report on Rio Grande do Sul for 1855, H. of C. 1859, session 2, XXX, p. 403.

14. Ibid., p. 404.

15. Jonathan C. Brown, *Socioeconomic History of Argentina*, p. 111.

16. See Gonçalves Chaves, *Memórias econômo-políticas*, pp. 116–18, 134–40, 141; Saint-Hilaire, *Viagem ao Rio Grande do Sul*, pp. 70–72.

17. V.S.L., art. 9.

18. Hunt, *Ure's Dictionary of Arts, Manufactures, and Mines*, 3:83; Jonathan C. Brown, *Socioeconomic History of Argentina*, p. 65.

19. Grigg, *Agricultural Revolution in South Lincolnshire*, especially pp. 148–49; Thompson, "Second Agricultural Revolution."

20. Inventory of deceased: Zeferina Gonçalves da Cunha; executor: Felisberto Inacio da Cunha, Barão de Correntes; Pelotas, 1886, N 1067, M 60, E 25, APRGS.

21. Maestri Filho, *Escravo no Rio Grande do Sul*, p. 91.

22. Dean, *Rio Claro*, p. 27.

23. Maestri Filho, *Escravo no Rio Grande do Sul*, pp. 91–92.

24. Ibid., p. 94.

25. Inventory of deceased: Antônio José Gonçalves Chaves; executor: João Maria Chaves; Pelotas, 1872, N 754, M 45, E 25, APRGS. This man was a son of the famous writer and charqueador of the same name who died in 1837.

26. V.S.L., art. 10.

27. See Correa da Câmara, "Ensaios statisticos do Rio Grande do Sul," pp. 22–23; Dreys, *Notícia descritiva da província do Rio Grande*, p. 132.

28. Fortunato Pimentel has argued that this became a route, chiefly in this century, for the diffusion of new ideas about ranching from Uruguay (*Aspectos gerais de Pelotas*, p. 81).

29. Inventory of deceased: João Simões Lopes; executor: João Simões Lopes Junior; Pelotas, 1853, N 366, M 26, E 25, APRGS.

30. The Fazenda Guaribú in Vassouras was valued at 434:721$000 in 1874; Stein, *Vassouras*, p. 247.

31. Mawe, *Travels in the Interior of Brazil*, p. 444. A similar argument can be found in Dreys, *Notícia descritiva da província do Rio Grande*, p. 132.

32. Jonathan C. Brown, *Socioeconomic History of Argentina*, p. 111.

33. Nicolau, "Industria saladeril en la confederacion argentina," pp. 23–24; Whigham, "Cattle Raising in the Argentine Northeast," pp. 325–26.

34. Dreys, *Notícia descritiva da província do Rio Grande*, p. 134.

35. Almeida was a leading administrator of the Riograndense Republic declared in 1836. He left extensive papers which form the core of the Alfredo Varela collection (Coleção Varela) of the AHRGS; CV 190, Domingos José de Almeida to Bernardina Barcelos de Almeida, Porto Alegre, 17 Feb. 1836, *AAHRGS* 2 (1978): 162–65.

36. The evidence lies in correspondence about his charqueada during the civil war; see, for example, CV 189, Almeida to Barcelos de Almeida, Porto Alegre, 16 Feb. 1836, and CV 545, Almeida to Luís Rodrigues Barcelos, Bagé, 25 Nov. 1841, *AAHRGS* 2 (1978): 160–62, 404–5.

37. Almeida owed the contemporary equivalent of over £17,000 to his creditors by 1831. CV 2173, Almeida to the respectable public and in particular to his friends ("Ao respeitável público e em particular aos meus amigos"), undated, incomplete document, *AAHRGS* 3 (1978): 642–44.

38. V.S.L., art. 21.

39. João Simões Lopes Neto maintained that Cascalho, a charqueada belonging to Domingos de Castro Antiqueira, Visconde de Jaguarí, was the first in Pelotas to manufacture charque using the "Platine system" and the only one to develop a specialized trade with Havana. This initiative was the result of efforts made by João Baptista Roux, a Frenchman who rented Cascalho around 1846. The slaughterers were French Basques brought from Montevideo, in addition to Argentineans and Uruguayans, as well as some thirty African slaves rented along with the plant ("Noticia sobre a fundação das xarqueadas," p. 11).

40. Luccock, *Notes on Rio de Janeiro*, p. 174; Saint-Hilaire, *Viagem ao Rio Grande do Sul*, p. 203; Avé-Lallemant, *Viagem pela província do Rio Grande do Sul*, p. 104; Vereker, *British Shipmaster's Handbook to Rio Grande do Sul*, pp. 3–4.

41. Vereker, *British Shipmaster's Handbook to Rio Grande do Sul*, p. 34.

42. At mid-century the port of Liverpool alone was accepting some half-million hides a year from South America. At least some factors engaged in the trade identified the hides by their general area of origin; thus, in 1837 a

Philadelphia merchant offered for sale "4,200 La Plata hides, 3,000 Chile hides, 1,000 Rio Grande hides, 800 La Guyra [La Guaira] hides, and 500 Pernambuco hides." See Jonathan C. Brown, *Socioeconomic History of Argentina*, pp. 58, 251 n 30.

43. A brief comment in one of the many editions of a dictionary of industrial processes hints that English tanners may have generally avoided sun-dried hides from South America due to their soaking time: "Hides imported from foreign countries, and which have been preserved by salting or drying, and *especially the latter*, require soaking for a longer period, in order to render them supple." Hunt, *Ure's Dictionary of Arts, Manufactures, and Mines*, 3: 82.

44. Simões Lopes Neto, "Noticia sobre a fundação das xarqueadas," p. 12.

45. Even for Rioplatense pastoral products, the British market was sensitive: "If the British based merchant could not comment on the depressed state of the market, then he was almost sure to remark on the poor condition of the produce sent. Jerked beef was improperly cured; hides were damaged by worms and had to be sorted or were the wrong weight for the market." Reber, *British Mercantile Houses in Buenos Aires*, p. 72.

46. Just why the American market looked for lighter hides remains obscure but probably relates to tanning methods. Hides from steers found their major use as soles for footwear, and tanning sole-leather took a year or more. Lighter hides were sometimes treated like calfskins (used for uppers), which only took three months to tan. See Hunt, *Ure's Dictionary of Arts, Manufactures, and Mines*, 3: 84–85. Using the prices prevailing for hides at New York and Liverpool in October 1859, deputy José d'Ávila da Silveira Amaro demonstrated to the provincial assembly of Rio Grande do Sul that Riograndense hides sold for lower prices than those from the Plata. In addition, merchants had six months' grace to pay for Riograndense hides, while those drawn from the Plata were paid for on sight. Speech by deputy Amaro, as reported in the *Correio do Sul*, 21 Jan. 1860.

47. See Jonathan C. Brown, *Socioeconomic History of Argentina*, p. 65.

48. Thompson, "Second Agricultural Revolution," p. 75; Vereker's report on Rio Grande do Sul for 1855, H. of C. 1859, session 2, XXX, p. 404.

49. Jonathan C. Brown, *Socioeconomic History of Argentina*, pp. 65–66; Vereker's report on Rio Grande do Sul for 1860, H. of C. 1862, LVIII, p. 646.

50. Vereker's report on Rio Grande do Sul for 1856, H. of C. 1859, session 2, XXX, p. 411; Vereker's report on Rio Grande do Sul for 1861, H. of C. 1863, LXX, p. 40.

51. V.S.L., art. 32.

52. Pedro de Alcântara Bellegarde, "Apontamento sobre a República do Paraguai e sobre a Província do Rio Grande do Sul," Asunción, 3 June 1849, manuscripts, 7, 4, 59, BN.

53. Galloway, *Sugar Cane Industry*, pp. 162–68.

54. Ryan, *Fish out of Water*, pp. 76–83, 205–25.

55. For an interpretation of why Rio de Janeiro's coffee plantations grew less subsistence food by the 1850's, see Stein, *Vassouras*, pp. 47–48.

56. Ryan, *Fish out of Water*, p. 213.

57. The best year for Rio Grande do Sul's charque exports during the

nineteenth century was 1868, when 45,458 tons were exported beyond the province, a quantity not surpassed until 1906. Newfoundland's worst year for salt-fish exports to Brazil during the period 1857–1914 was also 1868, when barely 5,000 tons were exported. Elmar Manique da Silva, "Ligações externas da economia gaúcha," p. 75; Ryan, *Fish out of Water*, p. 213.

58. See my "Early Industrialization in the South Atlantic," p. 401; Kerst, "Die brasilische Provinz Rio Grande do Sul," pp. 112–13.

59. Luccock, *Notes on Rio de Janeiro*, pp. 177, 182; Saint-Hilaire, *Viagem ao Rio Grande do Sul*, p. 62; Isabelle, *Voyage a Buénos-Ayres et a Porto-Alègre*, pp. 375–76, 600.

60. On these issues, see Gonçalves Chaves, *Memórias ecônomo-políticas*, especially pp. 93–99, 172, 192, 210–11.

61. The criticisms made by Gonçalves Chaves found a germ of official support in General Andréa's mid-century administration of Rio Grande do Sul. See Province of RGS, *Relatório, 1°.-06-1849*, pp. 10–11, and *Relatório, 6-3-1850*, pp. 16–17; Renato Costa, "Um precursor da reforma agrária," *Correio do Povo*, 11 May 1968.

62. See Lynch, "River Plate Republics," p. 649.

63. See my "Early Industrialization in the South Atlantic," pp. 404–5, 408.

64. See the speeches made by Jacintho de Mendonça in the chamber of deputies, sessions of 27 July and 2 August 1853, as reported in O *Mercantil*, 1, 2, and 3 Sept. 1853.

65. Bell, "Early Industrialization in the South Atlantic," pp. 406–7.

66. Hide exports fell from 893,287 in 1851 to 617,441 in 1855. Vereker's report on Rio Grande do Sul for 1856, H. of C. 1859, session 2, XXX, p. 408.

67. Commercial Association of Rio Grande, 20 July 1858; correspondence of the presidents, maço 29, AHRGS.

68. Province of RGS, *Relatório, 5-11-1858*, p. 42.

69. The treaty was approved by the Uruguayan *cámaras* in July 1858 and ratified in Rio de Janeiro on 23 September of the same year. See Barrán and Nahum, *HRUM*, 1(1): 93.

70. "It is to be regretted that the Treaty [of 1857] . . . is to be in vigour for only four years, as the proprietors in this province are kept in uncertainty as to whether the renewal of the Treaty will not be refused at the lapse of that period in view of the agitation got up against it, or whether they will not, at least, obtain as compensation a special tariff and reduced scale of duties; they are thus kept from developing the agricultural riches which, with a little energy, are within their reach." Vereker's report on Rio Grande do Sul for 1858, H. of C. 1861, LXIII, p. 504.

71. The profits from this intermediate trade concentrated in the hands of a small nucleus of merchants, of whom the immense majority were foreign and who controlled the greater part of the money in circulation in Uruguay. Barrán, *Apogeo y crisis del Uruguay*, pp. 61–62. On the factors that made Montevideo "a commercial centre of the first importance," see also Oddone, "Formation of Modern Uruguay," p. 454.

72. Analyzing the markets for salt-beef in Brazil and Cuba in 1862, Barrán and Nahum conclude that Uruguay contributed 37.4 percent of the animals

killed, Argentina 35.7 percent, and Rio Grande do Sul 26.8 percent. Argentina would probably have made a larger contribution, except that it was the scene of a major political conflict at that time between Justo José de Urquiza and Bartolomé Mitre. These statistics demonstrate well that Rio Grande do Sul was normally in a peripheral position as a supplier of salt-beef (*HRUM*, 1(1): 121–22).

73. *AALRGS*, session of 18 Sept. 1862, p. 25.

74. See the minute prepared by the provincial treasury; correspondence of the presidents, maço 33, AHRGS. Only a portion of this document has survived, so that fuller bibliographical details are unavailable.

75. Speech by deputy Nascimento in *AALRGS*, session of 18 Sept. 1862, p. 26.

76. They were not alone in this. The Uruguayans were also looking toward Europe as a market for beef, particularly to England. See the exposition on this subject by the board of directors of Uruguay's Club Nacional of 20 August 1862, as published in Barrán and Nahum, *HRUM*, 1(2): 31–38; Marriner, *Rathbones of Liverpool*, pp. 44–46.

77. Shortage of meat was already being felt in Europe, especially in France, where the taxes on imported salt-beef from the Plata and Brazil had been lowered in September 1853 and where formal interest was being expressed in the possibilities of using meat in conserved form. Province of RGS, *Relatório*, *2–10–1854*, pp. 45–46.

Chapter 4

1. Walton, "Diffusion of the Improved Shorthorn," p. 32.

2. Halperín-Donghi, *Politics, Economics and Society in Argentina*, p. 27.

3. Uricoechea, *Patrimonial Foundations of the Brazilian Bureaucratic State*, p. 142.

4. Duncan Baretta, "Political Violence and Regime Change," p. 222.

5. See Pebayle, *Les Gauchos du Brésil*, pp. 174–75.

6. See Duncan Baretta, "Political Violence and Regime Change," especially pp. 61–62, 207.

7. Pesavento, *República velha gaúcha*, p. 53.

8. Abente, "Foreign Capital, Economic Elites and the State in Paraguay," p. 81.

9. Slatta, *Gauchos and the Vanishing Frontier*, p. 94.

10. See the work treating the historical antecedents of the Asociación Rural del Uruguay, donated by Adolfo Linardi during 1981 to Uruguay's Museo Histórico Nacional in Montevideo, MHN, doc. 3563, pp. 49–52.

11. These works are extremely difficult to find, even in Brazil. Some of the articles in Cabral's *O Propagador da Indústria Rio-Grandense* were reprinted in newspapers and journals in other parts of the country. He was called away from his useful editorship in order to gather statistics for the provincial government of Rio Grande do Sul. For bibliographical details, as well as an account of Cabral's career in Brazil, see Barreto, *Bibliografia sul-riograndense*, 1: 233–41.

12. Slatta, *Gauchos and the Vanishing Frontier*, pp. 98–99.

13. There is an excellent brief synthesis of the circumstances of the formation of the Rural Association, as well as of its early program, in Reyes Abadie and Vázquez Romero, *Logros de la modernización*, pp. 419–27.

14. Barrán and Nahum, *HRUM*, 1(1): 88–90.

15. Oddone, "Formation of Modern Uruguay," p. 455.

16. See Barrán and Nahum, *HRUM*, 1(1): 330–31 and 1(2): 133–34.

17. Ibid., 1(1): 325–26, 358.

18. Quoted ibid., 1(1): 331.

19. Ibid., 1(1): 358.

20. Ibid., 1(1): 335.

21. Barrán and Nahum, *Un diálogo difícil*, pp. 254–82.

22. See Dean, "Green Wave of Coffee," especially pp. 94, 113. In 1881 the descendants of Eliseu Maciel asked the council at Pelotas for permission to construct an agricultural and veterinary school in honor of their ancestor. This project was well received. The province was to assume the cost of paying for teachers, but while the building was ready in 1883, there was difficulty contracting with staff. In the same year, the imperial government announced that it intended to create a school of agronomy and veterinary science in Rio Grande do Sul and that the Eliseu Maciel buildings could be used. A French veterinary surgeon, Claude Marie Rebourgeon, was sent to Pelotas to verify the suitability of the site. He reported favorably and sent various types of equipment, including laboratory materials and pedigree animals, from Europe to Pelotas. However, the Cotegipe government changed its mind about funding an institution in Pelotas. Seeking to transfer funds in order to found the Instituto Agronômico de Campinas, it ordered the equipment in Pelotas sold. Nevertheless, much of the material remained in Pelotas under the care of the Visconde da Graça, and in 1887 the council resolved to bring the school into being under the name of Liceu Riograndense de Agronomia, Artes e Ofícios. See "Pelotas comemora os cem anos da Faculdade Eliseu Maciel," *Correio do Povo* rural supplement, 2 Dec. 1983.

23. See Pesavento, *República velha gaúcha*, pp. 52–66.

24. The membership list is printed in *A Estância* (June 1917): 135. No locations were given for six members. Only one member was listed as residing outside the state, at Buenos Aires.

25. On the problems of viewing the Campanha through the theories of functional or polarized regions, see Rogério Haesbaert da Costa, *Latifúndio e identidade regional*, especially pp. 16–17.

26. Argentina's leading rural-economy journals during the century 1830–1930 are listed in Halperín-Donghi, "Argentina," pp. 140–42.

27. MHN, doc. 3563, pp. 117–18, 143.

28. In 1859, the councillors at Alegrete acknowledged receipt of a pamphlet emanating from this organization entitled "Regeneração das raças cavallares do Imperio do Brasil, para serem distribuidos convenientemente pelos estancieiros . . ." CM of Alegrete, doc. 973, 22 Jan. 1859, AHRGS. In addition, the council received the *Revista Polytechnica* by Dr. Schmidt of Hamburg. CM of Alegrete, docs. 586, 18 Dec. 1849, and 778, AHRGS.

29. There is some limited evidence as to what was being read in the provin-

cial capital. In an appendix to his report for 1878, provincial president Francisco de Faria Lemos included details of the types of works consulted in the public library in Porto Alegre during the previous year. A total of 1,473 readers were registered at the institution. There had been only seven consultations for agriculture, equal to the number for statistics. There were fewer consultations only for the categories of didactic works (five) and religion (three). Such serious reading as was taking place in the region must have been largely confined to private libraries. Province of RGS, *Relatório, 10-2-1878.*

30. Around 1910 this organization was still provisionally based in the Boulevard Beauséjour, far removed from the lands of Brazil. For a list of the members in 1908–9, as well as a copy of the society's statutes, see Assis Brasil, *Cultura dos campos*, pp. 353–75.

31. Assis Brasil had served for part of the 1890's as Brazilian minister in Argentina, where he must have noticed the important roles of both sheep and cereals in the generation of rural wealth, even though his official correspondence is silent on the topic. The first edition of *Cultura dos campos* ran to 4,000 copies and was printed in Lisbon, where Assis Brasil was serving as a diplomat. By 1910, when *Cultura dos campos* emerged in its third edition from Paris, there were 20,000 copies published. The society was also responsible for another work organized by Assis Brasil, a translation of a manual on sheep breeding written by an Australian colonist (*Guia do criador de carneiros*).

32. On the history of attempts at agricultural organization through journals in Rio Grande do Sul, see the letter by Júlio H. Petersen in the *Correio do Povo*, rural supplement, 4 Nov. 1983. The first number of the *Revista Agrícola* appeared carrying the date 31 July 1897. Petersen affirms that this journal was definitely published until the end of 1898. In fact, it continued to appear (with gaps) during the early years of the twentieth century, and an attempt was made to resurrect the publication in 1929.

33. This was partly because many of the best articles were written by Guillaume (Guilherme) Minssen, a French-born and -educated agronomist hired by the Maciel agricultural school of Pelotas in 1889. See Barreto, *Bibliografia sul-riograndense*, 1: 936–37.

34. *RARGS*, no. 1 (31 July 1897): 2–4. Agronomists in São Paulo showed similar concerns during this period about the lack of integration of cattle raising within agrarian systems; see Dean, "Green Wave of Coffee," p. 105.

35. *RARGS*, no. 1 (31 July 1901): 1.

36. *RARGS*, no. 1 (15 Jan. 1929): 1.

37. See "O nosso programma," *A Estância* (Mar. 1913): 4–5.

38. *A Estância* (Mar. 1913): 22–26.

39. See *Revista do 1° Centenario de Pelotas*, no. 2 (25 Nov. 1911): 3. *La Hacienda: Revista Mensual Ilustrada sobre Agricultura, Ganadería e Industrias Ruraes* was published by La Hacienda Company in Buffalo, New York. This title first appeared in 1905.

40. Estanislao Zeballos, quoted without source, in Gibson, "Evolution of Live-stock Breeding in the Argentine," p. 78.

41. Barrán and Nahum, *HRUM*, 1(1): 88.

42. Gibson, "Evolution of Live-stock Breeding in the Argentine," pp. 78–79.

43. Quoted in *Times* (London), 8 Feb. 1865.
44. See Emília Viotti da Costa, "O escravo na grande lavoura," pp. 169–71.
45. Stein, *Vassouras*, pp. 122–24.
46. Salvador Duarte, "Feira pastoril," *RAF*, no. 23 (9 Jan. 1909): 353–54; Pesavento, *República velha gaúcha*, pp. 54–55; Reverbel, *Capitão da Guarda Nacional*, pp. 183–87.
47. Monte Domecq et Cie., *Estado do Rio Grande do Sul*, p. 226.
48. Unlike the Plata, Brazil usually referred to the Shorthorn as the Durham. Since Shorthorn is the more widely understood term for the breed, it is the one used in the rest of this work.
49. *AALRGS* (1917): 42.
50. See the interview with Assis Brasil on the occasion of Porto Alegre's fourth cattle fair in the *Correio do Povo*, 23 Sept. 1920.
51. *RARGS*, no. 5 (31 Jan. 1902): 116.
52. "Resumo das operações da exposição feira realisada em 15-11-18, em São Borja," *A Estância* (Nov.–Dec. 1918): 269.
53. Slatta, *Gauchos and the Vanishing Frontier*, p. 107.
54. Ibid., p. 114.
55. See ibid., pp. 114–15.
56. Oddone, "Formation of Modern Uruguay," p. 458.
57. See, for example, the section "Costeio das Fazendas e registo das tropas" in the "código de posturas" (legal code) of Alegrete. CM of Alegrete, doc. 556, 9 June 1849, AHRGS.
58. *RARGS*, no. 4 (31 Oct. 1901): 60.
59. See the editorial "Considerações justas," *RAF*, no. 39 (30 Nov. 1909): 1–3.
60. *A Estância* (May 1913): 83–84.

Chapter 5

1. V.S.L., art. 9; *RAF*, no. 41 (1 Jan. 1910): 1–3.
2. See my "Aimé Bonpland and Merinomania," p. 301. Parts of the discussion on sheep in this chapter draw from this article.
3. See Crossley and Greenhill, "River Plate Beef Trade," p. 293.
4. For a fascinating analysis of changing human-animal interactions in the nineteenth century, see Ritvo, *Animal Estate*, especially pp. 45–81; see also Walton, "Pedigree and the National Herd." Both these studies show clearly how issues of social status confused utilitarian concerns.
5. Saint-Hilaire, *Viagem ao Rio Grande do Sul*, p. 46; V.S.L., art. 34. For a more optimistic report, see Gonçalves Chaves, *Memórias ecônomo-políticas*, pp. 202–4.
6. Cesar, *Conde de Piratini*, pp. 20, 41. See also the details of the advice that Rosas handed out when administrator of some of the Anchorena family properties in Argentina; Jonathan C. Brown, *Socioeconomic History of Argentina*, p. 137.
7. CM of Bagé, doc. 197, art. 8 (circular no. 14 of 21 July 1854); CM of Pelotas, doc. 378a, art. 8, 21 Aug. 1854; CM of Alegrete, doc. 936, 10 Jan. 1858, AHRGS. However, by 1868 the councillors of Alegrete referred to a

constant effort to improve the quality of the livestock. CM of Alegrete, doc. 1097, 12 Feb. 1868, AHRGS.

8. There is reference to 300 merino sheep and an English ram in inventory of deceased: Zeferino Maximano Ribeiro; executrix: Eugenia das Chagas Ribeiro; Bagé, 1859, N 149, M 7, E 38, APRGS; to improved horses in inventory of deceased: José Nogueira de Andrade; executrix: Felisbina Maria de Moura; Cruz Alta, 1864, N 95, M 4, E 61, APRGS; and to a small bull of improved breed ("tourinho de raça fina") in inventory of deceased: Antônio José Gonçalves Chaves; executor: João Maria Chaves; Pelotas, 1872, N 754, M 45, E 25, APRGS.

9. Province of RGS, *Relatório, 1º.-10-1852*, p. 33.

10. Province of RGS, *Relatório, 2-10-1854*, p. 45.

11. See the section "Cultura e industria dos diversos municipios da Provincia, mercados para onde são conduzidos os productos, o modo de transporte, e as obras que são por aquelles reclamadas como mais urgentes," ibid., p. 53.

12. There are numerous specific expressions of his interest in breeding in his correspondence. See the letters from General Andréa to the Barão de Jacuí, 10 May, 12 May, and 12 Aug. 1848, and from General Andréa to Commander Manoel Ferreira Porto Filho, 19 Sept. 1848; diverse documents, *caixa* 18, AHRGS. See also the letter from the publisher Firmin Didot to the provincial president, Rio de Janeiro, 20 Feb. 1850; correspondence of the presidents, maço 21, AHRGS.

13. See RGS, provincial law no. 131, established in the first session of the third legislature, 12 July 1848. This laid out a detailed system for breeding in the province which would have been a mixture of official and private enterprise.

14. Province of RGS, *Relatório, 1º.-06-1849*, pp. 28–29.

15. Province of RGS, *Relatório, 1º.-10-1852*, p. 33, and *Relatório, 2-10-1854*, p. 45.

16. Rock, *Argentina, 1516–1987*, pp. 133–34.

17. Barrán and Nahum, *HRUM*, 1(1): 33, 200; 1(2): 12.

18. McColl, *Republic of Uruguay*; Murray, *Travels in Uruguay*, especially pp. 155–234.

19. Province of RGS, *Relatório, 1º.-10-1852*, p. 33. Schmidt was the author of a work on German emigration which Abeillard Barreto considered a panegyric on southern Brazil. It was probably on the strength of this that he obtained his post from the Brazilian government. Of the group of expediting agents ("agentes expeditores") based in Hamburg, Schmidt was the one who specialized in shipments to Brazil. See Barreto, *Bibliografia sul-riograndense*, 2: 1229–30.

20. See the invoice for four bundles of hay dated in Hamburg on 15 Aug. 1853; correspondence of the presidents, maço 24, AHRGS.

21. Province of RGS, *Relatório, 6-10-1853*, p. 41.

22. Province of RGS, *Relatório, 2-10-1854*, p. 55.

23. According to Oliveira Belo, a special building was under construction, the work directed by Friedrich Heydtmann, one of the provincial engineers. See

Province of RGS, *Relatório, 27-9-1855,* p. 23. In fact, Phillip von Normann, another German engineer, played a major role in the construction of the breeding station. In the period between the post-Farroupilha pacification and the early stirrings of war with Paraguay, most of those involved in what we might broadly define as "engineering projects" in Rio Grande do Sul were of German origin. See the interesting article by Weimer, "Engenheiros alemães no Rio Grande do Sul," pp. 155, 183.

24. Province of RGS, *Relatório, 28-4-1856,* pp. 39–40.

25. Province of RGS, *Relatório, 11-10-1857,* pp. 26–27.

26. Province of RGS, *Relatório, 5-11-1858,* p. 34.

27. These statistics were incomplete, not least since the counties of Triunfo, Rio Pardo, Cachoeira, São Gabriel, Livramento, Uruguaiana, Itaqui, Jaguarão, and Rio Grande provided no data. Province of RGS, *Relatório, 5-11-1860,* p. 51.

28. See Rock, *Argentina, 1516–1987,* p. 134.

29. Province of RGS, *Relatório, 10-3-1864,* pp. 56–57.

30. On the general character of this gulf, see Barman, *Brazil,* p. 218.

31. Province of RGS, *Relatório, 27-9-1855,* p. 23.

32. See Bonpland, *Journal voyage de Sn. Borja a la cierra y a Porto Alegre,* pp. 66, 120, 131, 132.

33. Alfred Duclos to Sinimbu, Estância do Leão, 22 July 1853; correspondence of the presidents, maço 24, AHRGS.

34. Aimé Jacques Goujaud (1773–1858), better known as Bonpland, led an extraordinarily diverse life in South America. For a detailed account of his work with merinos, see my "Aimé Bonpland and Merinomania," especially pp. 308–12 and 315–23.

35. See the collection of letters from Bonpland to Antônio Rodrigues Chaves; manuscripts, I-2, 3, 36, BN.

36. Rock, *Argentina, 1516–1987 ,* p. 134; Scobie, *Argentina,* pp. 84–85.

37. Conde d'Eu, *Viagem militar ao Rio Grande do Sul,* pp. 36, 43.

38. *A Reforma,* 12 Oct. 1880.

39. See the letter from the president of Rio Grande do Sul to the Barão de Ijuí of 8 Oct. 1880, as reported in *A Reforma,* 12 Oct. 1880.

40. "Glorias rio-grandenses: uma 'cabaña' notavel," *RAF,* no. 45 (1 May 1910): 1. Another author ascribes the first importation of a herd of purebred Shorthorns to Dr. Gervasio Alves Pereira in 1895; Carvalho, *Nobiliário sul-rio-grandense,* p. 259.

41. *A Reforma,* 28 July, 6 Aug., 25 Aug., and 27 Oct. 1880.

42. *A Reforma,* 25 May 1880.

43. See the speech by deputy Villa Nova, AALRGS, session of 19 Nov. 1885, pp. 48–49.

44. In their general consideration of breeding stations, the assembly discussed the imperial government's efforts in establishing the invernada at Saicã. Assis Brasil claimed the results there replicated those of the unsuccessful experiments with sheep. Adding the comment, "At the end of the world," another deputy implied that Saicã's isolation was also an obstacle. One of the main problems with the national lands in São Gabriel was that they were home

to plenty of blood-sucking worms (*sanguesugas*) as well as horses. See the speech made in the session of 2 Jan. 1888 by Borges Fortes, in connection with the third debate over project 204, sponsored by Assis Brasil, *AALRGS* (1887): 403–4.

45. See, for example, CM of São Gabriel, "Informações sobre o estado geral da agricultura e da indústria pastoril, serica e agrícola," 12 Apr. 1883, AHRGS.

46. In defending the value of crossing the creole cattle with Shorthorn bulls, Reyles noted in 1882 that he had sold animals to every department in Uruguay, as well as into Brazil and Entre Ríos. See Barrán and Nahum, *HRUM*, 1(1): 603–4; 1(2): 292, 294.

47. Mariani, "Comienzos del processo de mestización ganadera," p. 94.

48. CM of São Gabriel, "Relatório com que a Câmara Municipal da Cidade de S. Gabriel (Província de S. Pedro do Rio Grande do Sul) passou seu mandato . . . ," 7 Jan. 1887, AHRGS.

49. Frances, *Beyond the Argentine*, p. 62.

50. Mariani, "Comienzos del processo de mestización ganadera," p. 98. In 1889, based on the values of the properties declared by Brazilians, they owned 80 percent of the department of Artigas. See Mourat, "Inmigración y el crecimiento de la población del Uruguay," p. 9.

51. A listing of animals ordered from England during the first six months of 1920 includes only 30 bulls of the leading beef breeds in a total of 75 animals (cattle, sheep, and pigs). However, this list presumably includes only animals sought through the agency of the Breeders' Union. See "Quadro dos animaes encommendados na Inglaterra, neste ultimo semestre, por criadores rio-grandenses," *A Estância* (June 1920): 149.

52. "Some progress is being made in improving the breed of cattle, but not such as to make any marked impression. Some high bred cattle are imported from Germany for breeding purposes, but it is probable that a much greater number are introduced from the River Plate." Wigg's report on Rio Grande for 1912, H. of C. 1913, LXIX, p. 5.

53. Calculated from data in Faria, *Diccionario geographico, historico e estatistico*, p. 185.

54. "Glorias rio-grandenses," p. 1.

55. "Novos horizontes," *RAF*, no. 41 (1 Jan. 1910): 1–3.

56. Giberti, *Historia económica da la ganadería argentina*, pp. 173, 175; Barrán and Nahum, *HRUM*, 6: 72.

57. It was founded in Bagé by Leonardo Collares and transferred to Pelotas in 1924. Reverbel, ed., *Diário de Cecília de Assis Brasil*, pp. 70–71.

58. *A Estância* (Nov.-Dec. 1916): 365 and (Dec. 1917): 304.

59. Barrán and Nahum, *HRUM*, 6: 72.

60. "Glorias rio-grandenses," pp. 1–8.

61. In 1919, an author who signed himself "Aldebarau" described the introduction of the Zebu to the "excellent pastures" of counties such as Bagé, Dom Pedrito, and Quaraí as a "crime." *Correio do Povo*, 6 Feb. 1919. There are many comments about the Zebu in *A Estância*, almost all of them critical.

62. Winsberg, "Introduction and Diffusion of the Aberdeen Angus in Argentina," pp. 188–89.

63. *RAF*, no. 48 (1 Aug. 1910): 8.

64. D. M. Riet, "A pecuaria riograndense," *O Criador Paulista* (Nov.-Dec. 1914): 1984. Although blind from the age of 21, Riet became one of Rio Grande do Sul's leading experts on the zootechnical aspects of ranching. He owned ranches in both Rio Grande do Sul and Uruguay. Most of Riet's articles are signed from Salto, Uruguay. See also his *O cavalo crioulo* and *Estância moderna*. For details of Polled Angus imports from Argentina, see *A Estância* (Nov. 1915): 784–86.

65. On his ranching experiences during the late Empire, see II Congresso Nacional da Agricultura, *Conferencia feita em sessão de 20 de agosto pelo Dr. Assis Brasil*, especially p. 13, and Assis Brasil, "Raças bovinas; qual a mais conveniente," *A Estância* (Sept. 1913): 207–12. Assis Brasil managed to observe agricultural conditions at an international level. While Brazilian minister in the United States, he attended the ranching fair held at Chicago in December 1900. He tried to solicit representation from Brazil, stressing the practical aspects of the gathering, but was not successful in this. For his impressions of the Chicago event, see *RARGS* (31 Dec. 1901): 81–84 and (31 Jan. 1902): 99–103. Later in his career, Assis Brasil visited the Royal Farm at Windsor. After that trip he supposedly quipped that he was unsure whether he had learned more from forty minutes at the farm or five years at the law school in São Paulo. Castilhos Goycochêa, *Eduardo de Araújo, Assis Brasil, Victor Russomano*, p. 84.

66. Pimentel, *Joaquim Francisco de Assis Brasil*, p. 63.

67. Polemical discussions about breeds can take decades to prove themselves. It is not clear how much science underlay Assis Brasil's arguments about breed selection, but he may have shown considerable foresight. South American ranching was a type of mining, continually drawing phosphates and lime from the land to promote animal growth but not replacing them, instead exporting them as animal products. For decades, South American animal bones and the phosphates they contained helped to fertilize British fields. This practice of exporting minerals appears to have created long-term problems for South American cattle breeders, at least in Uruguay. By the 1930's, ranchers there found evidence of osteomalacia, a cattle disease in which the loss of earthy salts causes softening of the bones. See Boerger, "Fundamental Factors of the Uruguayan Forage Problem," p. 76.

68. "Municipio de Alegrete," *A Estância* (Jan. 1917): 10.

69. "O Rio Grande observado pelos Argentinos," *Correio do Povo*, 4 Dec. 1920.

70. Pebayle, *Les Gauchos du Brésil*, p. 176.

71. Castilhos Goycochêa, *Eduardo de Araújo, Assis Brasil, Victor Russomano*, p. 35. From his daughter Cecília's diary, it appears that Assis Brasil could have made the claim by 1917. See Reverbel, ed., *Diário de Cecília de Assis Brasil*, p. 46. Assis Brasil considered the Devon the best regenerator of degenerated *mestiço* cattle bred in part from the great European breeds. In 1915, he imported the entire herd of the Cabaña Loraine into Rio Grande do Sul from

Uruguay. This was composed of 800 purebred Devons, and, although Assis Brasil said it himself, he was probably correct in asserting that there had never been anything of similar magnitude in relation to the progress of ranching in Brazilian history. See "Gado Devon," *A Estância* (May 1915).

72. Pedro Antonio Pereira de Souza, "Importancia da criação de ovelhas no Rio Grande do Sul," *A Estância* (Sept.-Oct. 1916): 254.

73. "Novos horizontes," *RAF*, no. 41 (1 Jan. 1910): 1–3.

74. D. M. Riet, "Criação de ovelhas," *A Estância* (July 1917): 166.

75. See Barrán and Nahum, *HRUM*, 7: 134–35.

76. D. M. Riet, "Criação de ovelhas," *A Estância* (July 1917): 166.

77. Intendência municipal de Jaguarão, *Apontamentos para uma monographia de Jaguarão*, especially p. 288.

78. See Reverbel, ed., *Diário de Cecília de Assis Brasil*, especially pp. 37, 42, 50.

79. "Ligeira noticia sobre a coudelaria e Fazenda Nacional de Saycan," *A Estância* (Jan. 1917): 6–8.

80. See Ordoñana's comments in Barrán and Nahum, *HRUM*, 1(2): 240.

81. Jacob, "Consecuencias sociales del alambramiento," pp. 123–24. For a general discussion of how wire fencing changed both pastoral landscapes and "genres de vie" in South America, see Deffontaines, "Contribution a une geographie pastorale de l'Amerique Latine," pp. 488–91.

82. For a description of a ranch with almost a kilometer of stone walls, see the inventory of deceased: Manoel Joaquim do Couto and Potenciana Joaquina do Couto; executor: Agostinho Pereira de Carvalho; Alegrete, 1848, N 92, M 7, E 65, APRGS.

83. Malin, *Grassland of North America*, p. 269; *Encyclopaedia Britannica*, 11th ed., s.v. "barbed wire."

84. Sbarra, *Historia del alambrado en la Argentina*, p. 94.

85. Joseph Love has noted that wire "began to spread across the campos of the Campanha and the Serra after 1870, and barbed wire followed in the 1880's" (*Rio Grande do Sul and Brazilian Regionalism*, p. 16).

86. Chasteen, *Heroes on Horseback*, p. 69.

87. Barrán and Nahum, *HRUM*, 1(1): 536.

88. Ibid., p. 537. Uruguayan estancias were generally divided into parcels of land (*potreros*) known as "suertes de estancia." Each suerte encompassed an area of 2,700 *cuadras cuadradas*, or a little short of 2,000 hectares.

89. Barrán and Nahum, *HRUM*, 1(1): 536–37.

90. Ibid., p. 544.

91. This bank was called the Caja de Crédito Territorial del Uruguay. See Charles Jones, "Commercial Banks and Mortgage Companies," p. 46.

92. Barrán and Nahum, *HRUM*, 1(1): 534–35.

93. Ibid., p. 542.

94. Ibid., p. 545.

95. Ibid., p. 543.

96. See Friedmann, *Urbanization, Planning, and National Development*, especially p. 51.

97. See the reports of the auxiliary commissions of the Rural Association for

the departments of Cerro Largo and Minas in *RARU*, no. 84 (1 June 1876): 165, 173–75.

98. See the speech made in the Brazilian senate on 4 October 1880, as reported in *A Pátria*, 26 Nov. 1880.

99. Bernal's report on Rio Grande do Sul for 1896, H. of C. 1897, LXXXIX, p. 509.

100. A Praça do Commercio da Cidade de Pelotas, doc. 1.

101. *A Pátria*, 1 Jan. 1881.

102. Lefebvre's report on Rio Grande do Sul for 1883, H. of C. 1884, LXXXI, p. 427.

103. Bennett's report on Rio Grande do Sul for 1885, H. of C. 1887, LXXXIII, p. 215.

104. Bernal's report on Rio Grande do Sul for 1897, H. of C. 1898, XCIV, p. 344.

105. Intendência municipal de Jaguarão, *Apontamentos para uma monographia de Jaguarão*, p. 297.

106. These accounts cover much of the last third of the nineteenth century. I am grateful to Sérgio da Costa Franco for providing me with a copy of them.

107. Province of RGS, *Relatório, 26-1-1879*, p. 27, and *Relatório, 19-7-1879*, p. 21.

108. CM of Bagé, doc. 580, 14 Apr. 1884, AHRGS. See especially art. 87.

109. CM of Bagé, doc. 617, 28 Jan. 1887, AHRGS.

110. José Antonio Moreira (1801–76) was granted the title of Barão de Butuí in 1873. He owned a charqueada in Pelotas which provided both a market for the fat cattle of, and some liquid capital for, the invernada at São João. Since Moreira's second wife was Leonidia Gonçalves, it seems safe to infer that he had a family connection with the Gonçalves da Silva family of Jaguarão. Carvalho, *Nobiliário sul-rio-grandense*, p. 39; inventory of deceased: José Antonio Moreira, Barão de Butuí; executor: Leopoldo Antunes Maciel; Pelotas, 1876, N 1834, M 88, E 26, APRGS.

111. He appears in the accounts under various spellings, but the phonetic rendering "Francisco Cheffar" leaves little doubt that his real surname was Schaefer, or some close variant of this. Summarizing the position of the Germans in Rio Grande do Sul in 1858, Avé-Lallemant recorded that he had found them on many of the estâncias, working as carpenters, workers, and shepherds. He also suggested that they were very well paid (*Viagem pela província do Rio Grande do Sul*, p. 377).

112. Among the expenses from October 1875 to 30 September 1876 there appears the following entry: "Sum paid to the German Francisco Cheffar from the contract which the invernada has with him to care for and replant the enclosure of the range."

113. There is a considerable variety in the situations described. As noted earlier, the Fazenda do Juncal received its first strands of Barbed wire in 1870. In 1912, União was still partly open land. Fencing, using ditches and banks, had begun in 1880 there, while the ranch built a sheep-dip "on the English system" in 1911. Barbed wire had been used to fence paddocks (potreiros), as seen in the statement that there existed "7 potreiros, with six and seven strands

of wire, with stone and wooden posts," divided into "3 invernadas, 1 for selection, another for bulls, another to *desterneirar* [separate the cows from their calves], and another for general use." Photographs of the Carolina ranch reveal that it was fenced at least in part with walls. It was also described as having nine potreiros with seven or eight strands of barbed wire. Finally, the Estância São Jeronymo was fenced "in part with mud-wall, *gravatá* [bromeliaceous vegetation] and wire and in part with six strands." See Intendência municipal de Jaguarão, *Apontamentos para uma monographia de Jaguarão*, especially pp. 287–88, 291, 297, 300.

114. For an interesting example from northern Uruguay, see Chasteen, *Heroes on Horseback*, pp. 99–100.

115. See Duncan Baretta, "Political Violence and Regime Change," especially p. 167.

116. *AALRGS*, session of 10 Mar. 1877, p. 22. Similar complaints were made about northern Uruguay; see the editorial "O codigo rural e os cercos" in *A Pátria*, 13 July 1880.

117. *AALRGS*, session of 16 Mar. 1877, pp. 55–56.

118. "Considerações justas," *RAF*, no. 39 (30 Nov. 1909): 1.

119. Vanger, *José Batlle y Ordoñez of Uruguay*, p. 5.

120. See Reyes Abadie and Vázquez Romero, *Logros de la modernización*, p. 432.

121. Some in Montevideo felt that those being displaced to Brazil ought to have been cultivating the land in agricultural colonies rather than leaving Uruguay. See *RARU*, no. 16 (31 Aug. 1886): 374–76.

122. Province of RGS, *Relatório, 19-7-1879*, p. 21.

123. Love, *Rio Grande do Sul and Brazilian Regionalism*, p. 72.

124. Part of this work includes a dubious sociobiological interpretation of the gaúchos, "strong types, as Ratzel would have said in lessons of anthropogeography." Beck, *Nova querência*, p. 23.

125. 50 quadras de sesmaria = 1 league.

126. RGS, "Relatório da Secretaria da Fazenda," Porto Alegre, 1909, quoted in Beck, *Nova querência*, p. 16.

127. See Barrán and Nahum, *HRUM*, 6: 388–89.

128. See, for example, Pfeifer, "Kontraste in Rio Grande do Sul," pp. 306–9; Roche, *Colonização alemã e o Rio Grande do Sul*, 1: 353–61.

129. Hirschman, *Development Projects Observed*, pp. 40–41.

130. "Cultivo dos pastos," *RAF*, no. 39 (30 Nov. 1909): 11.

131. Hinchliff, *South American Sketches*, p. 58.

132. Daireaux, "Estancias in Argentina," p. 4; Scobie, *Revolution on the Pampas*, p. 45.

133. Daireaux, "Estancias in Argentina," p. 13.

134. Boerger, "Grassland Panorama of the La Plata Region," p. 241.

135. Barrán and Nahum, *HRUM*, 6: 103–17.

136. Consul G. S. Lennon-Hunt could not understand why Rio Grande do Sul was importing alfalfa from Argentina "when the province [possessed] every advantage of soil and climate to grow it, not only for its own consumption but

for that of the whole of Brazil." Lennon-Hunt's report on Rio de Janeiro for 1873, H. of C. 1875, LXXV, p. 315.

137. Assis Brasil, *Cultura dos campos*, p. 321.

138. Ibid., pp. 197–99.

139. "Forragens," *RAF*, no. 23 (9 Jan. 1909): 354.

140. *RARGS*, no. 6 (31 Dec. 1897): 5.

141. Roche, *Colonização alemã e o Rio Grande do Sul*, 1: 250.

142. Clarence Jones, *South America*, pp. 364–65; Jefferson, "Pictures from Southern Brazil," pp. 530–31.

143. *AALRGS* (1917): 45. To place this area in perspective, by 1911 a single ranch in Buenos Aires Province was cultivating almost 17,000 hectares of lucerne grass, according to a British consul. See Mackie's report on Buenos Aires for 1911, H. of C. 1912–13, XCIV, pp. 126–27, quoted in Platt, *Latin America and British Trade*, p. 263.

144. Brazil, *Recenseamento de 1920*, 3(2): vi, xvi.

145. "Cultura de alfafa," *A Estância* (Apr. 1913): 68.

146. Boerger, "Fundamental Factors of the Uruguayan Forage Problem," p. 75, and "Grassland Panorama of the La Plata Region," p. 241.

147. Assis Brasil, *Cultura dos campos*, p. 302.

148. Scobie, *Revolution on the Pampas*, pp. 45–46.

149. Assis Brasil, *Cultura dos campos*, pp. 336–37.

150. Family diaries provide a clear picture of the effort entailed in natural pasture improvement. See, for example, Reverbel, ed., *Diário de Cecília de Assis Brasil*, p. 31, and Reverbel, *Pedras Altas*, pp. 17, 32, 43.

151. *Correio do Povo*, 21 Oct. 1919.

152. See the veterinary comments, including letters from Alegrete ranchers, in *Correio do Povo*, 7 June and 6 Nov. 1919.

153. See Reverbel, *Pedras Altas*, especially pp. 57–58, 65–66, 121.

154. *A Estância* (Nov. 1913): 268.

155. Pesavento, *República velha gaúcha*, pp. 119–20.

156. Dom Pedrito had 119 members and 11 tick-baths, Bagé 51 members and 41 tick-baths.

Chapter 6

1. Marriner, *Rathbones of Liverpool*, pp. 44–46.

2. In the early part of the twentieth century, Uruguay, for example, promoted the sale of salt-beef in Japan, Java, Sumatra, South Africa, and the Belgian Congo. See Barrán and Nahum, *HRUM*, 6: 155–59.

3. Gibson, "Evolution of Live-stock Breeding in the Argentine," pp. 92–93.

4. See Wade, *Chicago's Pride*, pp. 198–203.

5. *South American Journal*, 30 May 1885.

6. Grigg, *Agricultural Systems of the World*, p. 49.

7. Crossley and Greenhill, "River Plate Beef Trade," p. 287.

8. *AALRGS*, session of 18 Sept. 1862, p. 28.

9. *AALRGS*, session of 1 Oct. 1862, p. 131.

10. Nascimento provided an example of how commercial interests from Rio

Grande do Sul made economic progress out of the problems of the European poor. In Pelotas, a large candle and soap factory boiled up the shins of cattle for their oils. Seeing the hunger in England at first hand, the owner of this establishment organized a tin-plate factory and placed the residue from his other operation, "whose odor neither the noble deputy's sense of smell nor mine would be able to tolerate," into cans. Two spoons of the jelly in a bowl with a cracker made a worker's lunch. Such was the hunger in England, Nascimento observed, that even this product was used as food. *AALRGS*, session of 2 Oct. 1862, p. 141.

11. *AALRGS*, session of 13 Nov. 1862, pp. 153–54. The annals of the same session record a representation made by the provincial government (in the names of Manoel Lourenço do Nascimento, Felix da Cunha, and Gaspar Silveira Martins) to the Empire, asking that any successful efforts to send meat to Europe be exempted from export taxes. The petitioners noted that Uruguay had already made its first attempts at this speculative activity of introducing salted meats into Europe.

12. Scobie, *Revolution on the Pampas*, p. 43.

13. Province of RGS, *Relatório, 1°.-3-1863*, p. 65.

14. Province of RGS, *Relatório, 10-3-1864*, p. 63. See also annex F in the same document, the report of the provincial treasury, 18 Feb. 1864, II–III.

15. These figures are calculated from data in Elmar Manique da Silva, "Ligações externas da economia gaúcha," pp. 88–89.

16. Bakos, *RS: escravismo & abolição*, p. 33.

17. Ibid., p. 18.

18. Inventory of deceased: Francisco Martins da Cruz Jobim; executrix: Rita Zeferina de Oliveira Jobim; São Gabriel, 1880, N 324, M 18, E 107, APRGS.

19. Bakos, *RS: escravismo & abolição*, pp. 58–61.

20. See Maestri Filho, *Escravo no Rio Grande do Sul*, p. 75.

21. Inventory of deceased: Zeferina Gonçalves da Cunha; executor: Felisberto Inacio da Cunha, Barão de Correntes; Pelotas, 1886, N 1067, M 60, E 25, APRGS.

22. See the analysis of Gonçalves Chaves's ideas in Maestri Filho, *Escravo no Rio Grande do Sul*, pp. 80–85.

23. Couty, *Le maté et les conserves de viande*. There is a good discussion of Couty's conclusions in Maestri Filho, *Escravo no Rio Grande do Sul*, especially pp. 66, 85–89.

24. While there has been a consistent interest in slavery and the charqueadas, the matter of who performed the work of the slaughtering plants after abolition still demands detailed research. In the Campanha towns, peons displaced by the fencing of the ranches probably provided much of the labor force during the 1890's. By the civil war of 1923, some slaughtering plants sought labor from a much greater distance; one author maintains that Swift recruited White Russians, Lithuanians, and even Arabs to work in its Rosário do Sul charqueada. See Marques, *Episódios do ciclo do charque*, p.265.

25. Reyes Abadie and Vázquez Romero, *Población, comunicaciones y desarrollo urbano*, pp. 613–15.

26. *RARGS* (31 July 1901): 12.

27. Arquivo de Borges de Medeiros, doc. 4325, 22 Oct. 1902, published in *RIHGRGS* 125 (1989): 162–63.

28. Crossley and Greenhill, "River Plate Beef Trade," pp. 323–24.

29. Ibid., p. 332.

30. Barrán and Nahum, *HRUM*, 6: 159.

31. These rankings are based on the data presented in Pesavento, *República velha gaúcha*, p. 50.

32. These figures are calculated from data ibid., p. 37.

33. See Crossley and Greenhill, "River Plate Beef Trade," especially pp. 293–95.

34. See Joslin, *Century of Banking in Latin America*, pp. 39–41.

35. Barrán and Nahum, *HRUM*, 6: 168, 175.

36. *A Reforma*, 24 Feb. 1889.

37. *A Reforma*, 24 Mar. 1889.

38. Azambuja, *Anuário do Rio Grande do Sul para o ano de 1888*.

39. Ricardo Gutierrez, quoted in *A Reforma*, 6 Sept. 1889.

40. *A Reforma*, 26 Feb. 1889.

41. *A Reforma*, 2 May and 13 June 1889; Baguet, "Rio Grande-do-Sul," p. 390.

42. Love, *Rio Grande do Sul and Brazilian Regionalism*, p. 127; Crossley and Greenhill, "River Plate Beef Trade," pp. 293–94.

43. *A Reforma*, 7 Sept. 1889.

44. Hambloch's report on Rio de Janeiro for 1911–12, H. of C. 1913, LXIX, p. 14.

45. Pesavento, "República velha gaúcha: estado autoritário e economia," pp. 216–19.

46. See Love, *São Paulo in the Brazilian Federation*, pp. 225, 258.

47. See Topik, *Political Economy of the Brazilian State*, p. 100.

48. Downes, "Seeds of Influence," p. 92.

49. Hirschman, *Strategy of Economic Development*, pp. 184–85.

50. For Roosevelt's impressions, see Downes, "Seeds of Influence," p. 96; for Assis Brasil's, see Reverbel, *Pedras Altas,* pp. 133–34.

51. Downes gives a useful summary of the development of Farquhar's ranching schemes, but he is overly optimistic about their practical achievements ("Seeds of Influence," pp. 92–96).

52. By 1919, the markets for Brazilian frozen beef had broadened considerably, with most of the meat consumed by Italy, France, Britain, and Egypt (in that order of importance). See Hambloch's report on Brazil for 1919, H. of C. 1920, XLIII, pp. 7, 12.

53. "Of all the countries in the world, Brazil currently offers the best field of action for the frigoríficos." V. C. Gatherer, "Indústria frigorífica no Brasil," *O Criador Paulista* (Aug. 1915): 171–72.

54. On the roots of American interest in Brazilian beef, see Downes, "Seeds of Influence," especially p. 157.

55. "Possibilidades economicas da pecuaria no Brasil," *O Criador Paulista* (Apr. 1916): 105.

56. Barrán and Nahum, *HRUM*, 6: 206–8.

57. Pesavento, *República velha gaúcha*, p. 132.
58. Marques, *Episódios do ciclo do charque*, pp. 250–51.
59. Love, *Rio Grande do Sul and Brazilian Regionalism*, pp. 177–78, 187.
60. Pesavento, *República velha gaúcha*, pp. 145–47.

Chapter 7

1. For a discussion of "spread effects," see Myrdal, *Economic Theory and Underdeveloped Regions*, especially pp. 31–33.
2. See Rogério Haesbaert da Costa, *Latifúndio e identidade regional*, pp. 69–70.
3. Hirschman, *Development Projects Observed*, pp. 150–51.
4. See Jarvis, "Predicting the Diffusion of Improved Pastures in Uruguay," p. 498.
5. Burns, *Poverty of Progress*, p. 88; see also Chasteen, *Heroes on Horseback*, especially p. 142.
6. Burns, *Poverty of Progress*, p. 89.
7. Ibid., pp. 90, 151.
8. I have drawn from some of Sanford Mosk's phrasing here ("Latin America and the World Economy," p. 69).
9. Saint-Hilaire, *Viagem ao Rio Grande do Sul*, p. 193.
10. Geraldo de Faria Corrêa was a medical doctor, journalist, and poet. As the author of an article on moral education, published in the journal of a literary society (the Partenon Literário of Porto Alegre) in 1879, Faria Corrêa clearly expressed his concern that the council had a duty to promote economic development in the region.
11. CM of São Gabriel, "Relatório com que a Câmara Municipal da Cidade de S. Gabriel (Província de S. Pedro do Rio Grande do Sul) passou seu mandato em 7 de janeiro de 1887," AHRGS.
12. Frances, *Beyond the Argentine*, p. 62.
13. CM of Bagé, doc. 573, 28 Mar. 1883, AHRGS.
14. CM of São Gabriel, "Informações sobre o estado geral da agricultura e da indústria pastoril, serica e agrícola," 12 Apr. 1883, AHRGS. The council left the crops unspecified.
15. Speech made by deputy Villa Nova, *AALRGS*, session of 19 Nov. 1885, pp. 48–49; CM of São Gabriel, "Relatório . . . de 1887," AHRGS.
16. On this theme, see Chasteen, *Heroes on Horseback*, especially pp. 68, 75–76, 143.
17. CM of São Gabriel, "Informações sobre o estado geral da agricultura," 12 Apr. 1883, AHRGS.
18. CM of Bagé, doc. 573, 28 Mar. 1883, AHRGS.
19. CM of São Gabriel, "Relatório . . . de 1887," AHRGS.
20. CM of São Gabriel, "Informações sobre o estado geral da agricultura," 12 Apr. 1883, AHRGS.
21. See Chasteen, *Heroes on Horseback*, p. 70.
22. CM of Bagé, doc. 627a, 9 Oct. 1888; *Relatório da Câmara Municipal de Bagé apresentado á Assembléa Legislativa Provincial* (Bagé, 1888), AHRGS.

23. Colônia Nova (see Figure 1.3), a Mennonite colony initially based on wheat farming, was founded near Bagé in 1949. Wheat was not a successful crop over the longer run. The landholdings were enlarged, and those colonists remaining switched to dairying. The colony's relative success in dairying was followed by the government-inspired Colônia Nova Esperança, set up in 1978 to settle some of the landless of the Planalto. Rogério Haesbaert da Costa, *Latifúndio e identidade regional*, pp. 68–70. Landlessness in the Serra is a major problem for present-day Rio Grande do Sul; the idea that some of the squatters in the Serra could be resettled in the Campanha remains an active political issue.

24. Dean, "Latifundia and Land Policy in Nineteenth-Century Brazil."

25. Oddone, "Formation of Modern Uruguay," p. 458.

26. *A Estância* (May 1913): 83–84. There is plenty of evidence that most of the "obligations" fell on those of limited means in Argentina and Uruguay.

27. Ibid., pp. 84–85.

28. Duncan Baretta and Markoff, "Civilization and Barbarism," pp. 592, 611.

29. Ibid., p. 603.

30. See the interesting discussion of informal contacts with nomadic labor ibid., p. 593.

31. See ibid., p. 599.

32. Quoted in Barrán and Nahum, *HRUM*, 1(1): 492.

33. Mulhall and Mulhall, *Handbook of the River Plate*, p. 623.

34. Quoted in Barrán and Nahum, *HRUM*, 1(1): 493.

35. Burns, *Poverty of Progress*, p. 124.

36. I do not agree with Chasteen's depiction of a world of "encroaching fences" as "static," but on the social pressures of this period, see his *Heroes on Horseback*, pp. 69–71.

37. Mulhall and Mulhall, *Handbook of the River Plate*, p. 628.

38. CM of Alegrete, doc. 1298a, 1877, AHRGS.

39. See the petition from the council at Jaguarão addressed to the provincial assembly and printed within a speech on rural crime delivered by deputy Diana. *AALRGS*, session of 20 Mar. 1877, pp. 16–20.

40. CM of Bagé, doc. 627a, 9 Oct. 1888, AHRGS.

41. See Burns, *Poverty of Progress*, p. 127.

42. The 1890 census gave the Campanha's population as 224,195. See Love, *Rio Grande do Sul and Brazilian Regionalism*, pp. 72, 130.

43. Pesavento, *República velha gaúcha*, p. 136.

44. *A Estância* (Mar. 1915): 480.

45. Duncan Baretta and Markoff, "Civilization and Barbarism," p. 620.

46. Gerald S. Graham, "Ascendancy of the Sailing Ship," p. 81; Gollan's report on Rio Grande do Sul and Santa Catarina for 1878, H. of C. 1878–79, LXXI, p. 369.

47. On the poor state of the port at Buenos Aires before the improvements, see Reber, *British Mercantile Houses in Buenos Aires*, p. 76, as well as the graphic description in Murray, *Travels in Uruguay*, pp. 51–53.

48. Compare, for example, Gollan's report on Rio Grande do Sul for 1863, H. of C. 1865, LIII, p. 370, with Munro's report on Montevideo for 1872, H. of C. 1873, LXV, p. 582.

49. See the provincial assembly debates of December 1859, especially deputy Amaro's speeches, as reported in the *Correio do Sul*, 21 and 22 Jan. 1860.

50. For further details of how the poor state of the bar was impeding regional economic progress during 1881, see Province of RGS, *Relatório, 14-1-1882*, pp. 27–31.

51. These freight rates are drawn from articles on transport published in the *Gazeta de Porto Alegre*, transcribed in *A Pátria*, 13 and 15 Oct. 1880. It appears that with the difficulty of the bar, freight rates there still approximated those common earlier in the century; see Oribe Stemmer, "Freight Rates Between Europe and South America," p. 24. Around 1880, Montevideo remained the best port of the region; in Argentina, the steamers of Lamport & Holt went up the estuary to Rosario, instead of loading eighteen kilometers off land at Buenos Aires; see Egerton's report on the Argentine Republic for 1879, H. of C. 1881, LXXXIX, p. 120.

52. Platt, *Latin America and British Trade*, p. 119.

53. See Ridings, *Business Interest Groups in Nineteenth-Century Brazil*, p. 276.

54. Macedo, *Ingleses no Rio Grande do Sul*, pp. 64–66; letter from Sir John Hawkshaw to Dom Pedro II, London, 5 Aug. 1875; doc. 7886, Arquivo da Casa Imperial do Brasil, Petrópolis, quoted in Richard Graham, *Britain and the Onset of Modernization in Brazil*, p. 92.

55. See Scobie, *Buenos Aires*, pp. 70–91.

56. Reyes Abadie and Vázquez Romero, *Población, comunicaciones y desarrollo urbano*, pp. 618–20.

57. Topik, *Political Economy of the Brazilian State*, p. 66.

58. For further detail on Farquhar, see Gauld, *The Last Titan*; Topik, *Political Economy of the Brazilian State*, pp. 100–101, 103, 105. On Farquhar's efforts to speed regional development, see my "Foreign Investment and the Historical Geography of Brazil," especially pp. 93–103.

59. Wigg's report on Rio Grande for 1912, H. of C. 1913, LXIX, p. 4.

60. Oribe Stemmer, "Freight Rates Between Europe and South America," pp. 25–26, 45.

61. Hambloch's report on Brazil for 1919, H. of C. 1920, XLIII, p. 51.

62. See Love, *Rio Grande do Sul and Brazilian Regionalism*, p. 222.

63. Province of RGS, *Relatório, 4-8-1865*, pp. 28–29.

64. Gollan's report on Rio Grande do Sul for 1863, H. of C. 1865, LIII, p. 375.

65. Gollan's report on Rio Grande do Sul for 1878, H. of C. 1878–79, LXXI, p. 375.

66. Lewis, *British Railways in Argentina*, pp. 18–21.

67. See Castro, *Treatise on the South American Railways*.

68. Joslin, *Century of Banking in Latin America*, especially pp. 26, 34, 64, 67; on the conservatism of British banks, see also Ridings, *Business Interest Groups in Nineteenth-Century Brazil*, p. 136.

69. Marchant, *Viscount Mauá and the Empire of Brazil*, pp. 133–34.

70. *Diário do Rio Grande*, 10 Feb. 1860.

71. Charles Jones, "Commercial Banks and Mortgage Companies," pp. 47–48.

72. Topik, *Political Economy of the Brazilian State*, especially pp. 42, 52.

73. Charles Jones, "Commercial Banks and Mortgage Companies," pp. 51–52.

74. Topik, *Political Economy of the Brazilian State*, pp. 64–65, 73.

75. Pesavento, *República velha gaúcha*, p. 102.

76. See ibid., especially pp. 118–19.

77. On the deficiencies of the regional railways in this period, see Kliemann, "A ferrovia gaúcha." In 1919, when the railways of Rio Grande do Sul lacked rolling stock to transport animals to an exhibition in the central states of Brazil, it was pointed out that Paraguay was sending its cattle to the frigoríficos of Buenos Aires without difficulty, which brought the further comment: "In this respect, then, we are beneath Paraguay; there is no possible replica." *Correio do Povo*, 11 Apr. 1919.

78. Pesavento, *República velha gaúcha*, especially pp. 101, 119, 186–87.

79. Ibid., pp. 225–26.

80. See Antonacci, *RS: as oposições & a revolução de 1923*, especially pp. 61–62.

81. See Adelman, "Agricultural Credit in the Province of Buenos Aires" and *Frontier Development*, pp. 193–205. See also Solberg, *The Prairies and the Pampas*, pp. 142–44.

82. Hambloch's report on Brazil for 1919, H. of C. 1920, XLIII, p. 30.

83. See Downes, "Seeds of Influence," especially pp. 490–94.

84. Love, *Rio Grande do Sul and Brazilian Regionalism*, p. 223.

85. Ibid., p. 102.

86. See Scobie, *Buenos Aires*, especially p. 11.

87. For an early forecast of the different sources of future urban growth in Porto Alegre and Pelotas, see Province of RGS, *Relatório, 6-10-1853*, p. 36.

88. For a useful introduction to the role of minorities in the town of Rio Grande, see Copstein, "Trabalho estrangeiro no Rio Grande."

89. Conde d'Eu, *Viagem militar ao Rio Grande do Sul*, pp. 134–35.

90. In 1891, there were 42 British subjects registered at the British consulate in Rio Grande, as compared with 19 listed at Porto Alegre. Only two Britons were recorded as living in Pelotas, both of whom described themselves as merchants. Walter Risley Hearn, British consul, to Júlio de Castilhos, Rio Grande, 23 Sept. 1891; British consular correspondence, maço 11, AHRGS. Ridings's argument that "most directors of the Agricultural Industrial Center of Pelotas were probably foreign" is doubtful. The fact that the Luso-Brazilian elite sat as intermediaries between the rural producers and the foreign merchants helps to explain "their strange mix of conservative and radical attitudes" (*Business Interest Groups in Nineteenth-Century Brazil*, pp. 45, 219–20).

91. See Reverbel, *Capitão da Guarda Nacional*, pp. 79–82.

92. See A Praça do Commercio da Cidade de Pelotas, doc. 9.

93. *A Pátria*, 11 Nov. 1880.

94. A Praça do Commercio da Cidade de Pelotas, doc. 9, p. 9. In a period when slavery was coming increasingly into question, it is interesting that the Pelotas merchants still considered their town wealthy.

95. See the commentary from the *Echo da Fronteira* quoted in *A Reforma*, 12 Feb. 1880.

96. Friedmann, *Urbanization, Planning, and National Development*, especially pp. 46–47.

97. Bonpland, *Journal voyage de Sn. Borja a la cierra y a Porto Alegre*, p. 64.

98. See Copstein, "Trabalho estrangeiro no Rio Grande," pp. 23, 30.

99. There is a good discussion of this topic in Reber, *British Mercantile Houses in Buenos Aires*, pp. 129–36.

100. Barrán and Nahum, *HRUM*, 1(1): 334–35 and 1(2): 135–36.

101. In 1910, the British owned just over 300,000 hectares of land in Uruguay, 70 percent of which was in the departments of Rio Negro and Paysandú. Kennedy's report on Uruguay for 1910, H. of C. 1911, XCVII, p. 18; Dunlop's report on Uruguay for 1911, H. of C. 1912–13, CI, p. 4.

102. Reber, *British Mercantile Houses in Buenos Aires*, p. 132.

103. See *RARU*, no. 28 (1 Feb. 1874): 57–58.

104. I have adapted Gilberto Freyre's phrasing here (*The Mansions and the Shanties*, p. 20).

105. Assis Brasil, 27 June 1899; Washington, *ofícios recebidos*, 1899; 233, 4, 12, AHI.

106. Assis Brasil, *Granja de Pedras Altas*.

107. Downes, "Seeds of Influence," p. 30.

108. See, for example, the speech made by deputy Mathias Teixeira de Almeida in the provincial assembly, session of 23 Dec. 1859, as reported in the *Correio do Sul*, 27 Jan. 1860. For comparative comment on state institutions in Rio Grande do Sul and Uruguay, see Chasteen, *Heroes on Horseback*, pp. 57–58.

109. See the *Correio do Sul*, 23 May 1860.

110. Crossley and Greenhill, "River Plate Beef Trade," p. 316.

111. *O Commercial*, quoted in *A Pátria*, 20 Oct. 1880.

112. Duncan Baretta, "Political Violence and Regime Change," p. 31. Instead of lowering tariffs, Rui Barbosa, the minister of the treasury in the provisional government, attacked contraband by setting up special fiscal zones along the Uruguayan frontier during 1890. They were abolished as early as October 1891. In addition to these ineffectual measures, the first republican state government developed a taxation policy that implied a regional discrimination against the Campanha by placing its highest export taxes on charque and hides. See Love, *Rio Grande do Sul and Brazilian Regionalism*, p. 46.

113. See the speech by deputy Candido Gomes, as reported in the *Correio do Sul*, 20 Jan. 1860. See also that by deputy Teixeira de Almeida, session of 21 Dec. 1859, as reported in the same newspaper, 22 Jan. 1860.

114. Barrán, *Apogeo y crisis del Uruguay*, p. 68.

115. Province of RGS, *Relatório*, 10-2-1878, p. 31.

116. *AALRGS*, session of 22 Dec. 1883, pp. 258–59.

117. Love, *Rio Grande do Sul and Brazilian Regionalism*, p. 102.

118. See the petition from Bagé of 15 Sept. 1903, printed in *AALRGS*, session of 20 Nov. 1903, pp. 83–93.

119. See the speech by deputy Villa Nova, *AALRGS*, session of 19 Nov. 1885, pp. 48–49.

120. Quoted in V. M. Carrió, "Instalación de un frigorífico en Rio Grande do Sul," Porto Alegre, 25 Mar. 1915; correspondence of the Uruguayan Consulate, maço 25, AHRGS.

121. See Love, *Rio Grande do Sul and Brazilian Regionalism*, especially pp. 222–23.

Chapter 8

1. State of RGS, *Relatório de Estatística, 1915*, p. 128. The quantitative data in this cross-section are drawn mainly from Brazil's first agricultural census of 1920. All first efforts at taking a census have their defects, and Brazil's was no exception. For example, the organizers of the census were unsuccessful in their intentions to collect reliable material on purebred animals or the state of rural credit. Nor does the census offer much of value on the extent or quality of pasture. Where possible, the deficiencies of the census have been covered using statistics collected by the state government of Rio Grande do Sul.

2. For a valuable study of the process of land inheritance in Rio Grande do Sul, see Chasteen, "Background to Civil War."

3. These statistics on average property sizes have been computed from data in Brazil, *Recenseamento de 1920*, 3(2): 184–91.

4. Data on foreign ownership of land in Rio Grande do Sul can be found ibid., 3(1): xxxvii, 12–13. The total area of land owned by the British in Rio Grande do Sul in 1920 was less than 7 percent of the area they had owned in Uruguay in 1910. See Kennedy's report on Uruguay for 1910, H. of C. 1911, XCVII, p. 18. It remains unclear how many of the Uruguayans owning land in Rio Grande do Sul had Luso-Brazilian cultural roots.

5. Ildefonso Simões Lopes was a descendant of the well-known family of Pelotas charqueadores.

6. *Correio do Povo*, 24 Sept. 1919.

7. "O Rio Grande observado pelos Argentinos," ibid., 4 Dec. 1920. This account had been published initially as articles in the Buenos Aires newspaper *La Razón*, a place where the commentators had no pressing need to be diplomatic in their judgments.

8. Brazil, *Recenseamento de 1920*, 3(1): lxiii.

9. These figures are calculated from the data ibid., pp. 486–91.

10. In an essay on the regional contrasts within Rio Grande do Sul, Gottfried Pfeifer drew attention to the national significance of Rio Grande do Sul's commercial agriculture by the 1960's. As can be seen, the state was already important four decades earlier ("Kontraste in Rio Grande do Sul," pp. 296–97).

11. Brazil, *Recenseamento de 1920*, 3(2): xxxvi.

12. Ibid., 3(2): xxvi, and xxxiii.

13. Ibid., 3(3): viii–ix.

14. On the development of Erechim, see Roche, *Colonização alemã e o Rio Grande do Sul*, 1: 281–83.

15. See ibid., 1: 244.

16. Brazil, *Recenseamento de 1920*, 3(3): 176–78.

17. See *A Estância* (May 1919): 40.

18. Brazil, *Recenseamento de 1920*, 3(3): viii, 82–85.

19. I have used the equation of 4 head of sheep = 1 of cattle based on the regional sense of equivalence contemporary with the census; see *A Estância* (July 1917): 58–59. Since the 1920 census collected no data specifically on pasture, Figure 8.8 must be viewed as a rough approximation. The census takers treated pasture as a residual, presuming that it accounted for almost all the area left when the amounts in cultivation and in forest were deducted from the total area of rural establishments surveyed; for which land uses the census included, see Brazil, *Recenseamento de 1920*, 3(2): ix, xii.

20. *Correio do Povo*, 3 May 1919.

21. Fortunato Pimentel, "Os frigoríficos," ibid., 4 Feb. 1920.

22. Ibid., 21 May 1919.

23. "O Rio Grande observado pelos Argentinos," *Correio do Povo*, 4 Dec. 1920.

24. Rogério Haesbaert da Costa, *Latifúndio e identidade regional*, pp. 57, 62.

25. Ibid., p. 58.

26. See Benetti, "Agropecuária na região sul do Rio Grande do Sul," pp. 105, 185.

27. The 1988 municipal livestock census gives a herd of 591,562 cattle at Alegrete; see Fundação Instituto Brasileiro de Geografia e Estatística, *Regiões Sul e Centro-Oeste*, p. 141.

28. See the section "Campo & Lavoura" in *Zero Hora*, 1 Dec. 1989.

29. Ibid.

30. See Rogério Haesbaert da Costa, *Latifúndio e identidade regional*, especially pp. 59, 63.

31. Alonso, "Análise do crescimento da região sul," p. 65; on clandestine slaughter, see Benetti, "Agropecuária na região sul do Rio Grande do Sul," p. 114.

32. Benetti, "Agropecuária na região sul do Rio Grande do Sul," p. 131.

33. Kohlhepp, "Donauschwaben in Brasilien," pp. 376, 380, and "Strukturwandlungen in Nord-Paraná," pp. 5–53.

34. Waibel, "Princípios da colonização européia no sul do Brasil," p. 271.

Conclusion

1. See Friedmann, *Urbanization, Planning and National Development*, especially pp. 41–64.

2. See Barrán and Nahum, "Uruguayan Rural History," pp. 659–62.

3. Ibid., p. 668.

4. See Leff, *Underdevelopment and Development in Brazil*, especially 1: 146–55, 227–28, and vol. 2.

5. For the historical roots of this idea, see Street, *Artigas and the Emancipation of Uruguay*, especially pp. 57–61.

6. Leff, "Economic Development and Regional Inequality."

7. For a survey of contemporary Brazilian ideas about the country's human capital, see Skidmore, "Racial Ideas and Social Policy in Brazil." On the lack of attention to local labor in another frontier region, see Wilcox, "Paraguayans and the Making of the Brazilian Far West," pp. 499–500.

8. For interesting comment on the idea that neighboring countries can pose economic challenges which result in the geographical transformation of specific areas, see the foreword by John D. Wirth in Solberg, *The Prairies and the Pampas*, pp. xi–xii.

9. Lynch, *Argentine Dictator*, p. 309.

10. Patterns of change in Campanha ranching illustrate some of Leff's points about the causes of a low rate of technical progress in nineteenth-century Brazil (*Underdevelopment and Development in Brazil*, 1: 139–41).

11. See the recent analysis of ranchers' risk aversion strategies in Bandeira, "Raízes históricos do declínio da região sul," pp. 19–23.

12. Pesavento, *República velha gaúcha*, especially p. 290.

13. Kirby, "Uruguay and New Zealand," especially pp. 141–42.

14. See Benetti, "Agropecuária na região sul do Rio Grande do Sul," especially pp. 130–34.

15. See Rogério Haesbaert da Costa, *Latifúndio e identidade regional*, especially pp. 77–87.

16. Garavaglia and Gelman, "Rural History of the Río de la Plata," p. 93.

17. On the importance of regional approaches to ranching, Terry Jordan's recent important study is extremely suggestive (*North American Cattle-Ranching Frontiers*, especially pp. 308–14).

18. Brookfield, *Interdependent Development*, p. 208.

19. For an interesting but brief study of a single rural enterprise, see Cabeda, "José Antônio Martins, pioneiro esquecido."

Bibliography

Archives

Arquivo Histórico do Itamarati
Arquivo Histórico do Rio Grande do Sul
Arquivo Público do Estado do Rio Grande do Sul
Biblioteca Nacional, Rio de Janeiro (seção de manuscritos)
Museo Histórico Nacional, Montevideo
Museu da Comunicação Social Hipólito José da Costa

Government Publications

Argentine Republic. *Agricultural and Pastoral Census of the Nation.* Vol. 3, *Stock-breeding and Agriculture in 1908: Monographs.* Buenos Aires: Printing Works of the Argentine Meteorological Office, 1909.

Brazil. *Relatorio da repartição dos negocios estrangeiros apresentado á assembléa geral legislativa na terceira sessão da oitava legislatura pelo respectivo ministro e secretario de estado, Paulino José Soares de Souza.* Rio de Janeiro: Typographia Universal de Laemmert, 1851.

———. *Recenseamento do Brazil realizado em 1 de setembro de 1920.* 5 vols. Rio de Janeiro, 1922–30.

Province of Rio Grande do Sul. *Relatório enviado à Assembléia Legislativa em 1°.-06-1849, pelo presidente Francisco J. Soares de Andréa; 2ª. sessão da 3ª. legislatura.*

———. *Relatório apresentado pelo presidente Francisco Soares de Andréa ao passar a administração ao presidente José Antônio Pimenta Bueno em 6-3-1850.*

———. *Relatório enviado à Assembléia Legislativa em 1°.-10-1852, pelo vice-presidente Luís Alves de Oliveira Bello; 1ª. sessão da 5ª. legislatura.*

———. *Relatório apresentado à Assembléia Legislativa em 6-10-1853, pelo presidente Cansansão Sinimbú; na 2ª. sessão da 5ª. legislatura.*

———. *Relatório apresentado à Assembléia Legislativa em 2-10-1854 pelo presidente Cansansão Sinimbú; 1ª. sessão da 6ª. legislatura.*

———. *Relatório apresentado ao barão de Muritiba pelo vice-presidente Luís Alves de Oliveira Bello, em 27-9-1855, na entrega da Presidência.*

———. *Relatório do barão de Muritiba na entrega da presidência a Jerônimo Francisco Coelho, em 28-4-1856.*

———. *Relatório do vice-presidente Patrício Correa da Câmara à Assembléia Legislativa em 11-10-1857, na 2ª. sessão da 7ª. legislatura.*

———. *Relatório apresentado à Assembléia Legislativa em 5-11-1858, pelo presidente Ângelo Moniz da Silva Ferraz; na 1ª. sessão da 8ª. legislatura.*

———. *Relatório apresentado à Assembléia Legislativa, em 5-11-1860, pelo presidente Joaquim Antão F. Leão; na 1ª. sessão da 9ª. legislatura.*

———. *Relatório apresentado à Assembléia Legislativa em 1º.-3-1863 pelo presidente Esperidião Barros Pimentel, na 2ª. sessão da 10a. legislatura.*

———. *Relatório apresentado à Assembléia Legislativa, em 10-3-1864 pelo presidente Esperidião Barros Pimentel, na 1ª. sessão da 11ª. legislatura.*

———. *Relatório apresentado pelo presidente Marcelino de Souza Gonzaga na entrega da presidência ao Visconde de Boa Vista em 4-8-1865.*

———. *Relatório apresentado pelo presidente Francisco de Faria Lemos ao entregar o governo ao vice-presidente João Chaves Campello, em 10-2-1878.*

———. *Relatório apresentado pelo presidente Américo de Moura Marcondes de Andrade ao passar a presidência ao Dr. Felisberto Pereira da Silva em 26-1-1879.*

———. *Relatório apresentado pelo presidente Felisberto Pereira da Silva ao entregar a presidência ao Dr. Carlos Thompson Flores, em 19-7-1879.*

———. *Relatório apresentado pelo presidente Francisco de Carvalho Soares Brandão ao entregar o governo ao vice-presidente Dr. Joaquim Pedro Soares, em 14-1-1882.*

State of Rio Grande do Sul. *Relatório da Repartição de Estatística: Secretaria dos Negócios do Interior e Exterior do Estado do Rio Grande do Sul, 1914.*

———. *Relatório da Repartição de Estatística: Secretaria dos Negócios do Interior e Exterior do Rio Grande do Sul, 1915.*

———. *Relatório da Repartição de Estatística. Secretaria dos Negócios do Interior e Exterior do Estado do Rio Grande do Sul, 1918.*

State of Rio Grande do Sul. Arquivo Histórico. *Guia do Acervo e inventário sumário dos códices do Arquivo Histórico do Rio Grande do Sul.* Caxias do Sul, 1980.

United Kingdom. House of Commons sessional papers (Parliamentary Papers) 1859, session 2, XXX; 1861, LXIII; 1862, LVIII; 1863, LXX; 1865, LIII; 1873, LXV; 1875, LXXV; 1878–79, LXXI; 1881, LXXXIX; 1884, LXXXI; 1887, LXXXIII and LXXXVI; 1893–94, XCVIII; 1897, LXXXIX; 1898, XCIV; 1911, XCVII; 1912–13, XCIV and CI; 1913, LXIX; 1920, XLIII.

Newspapers and Periodicals

Anais da Assembléia Legislativa do Rio Grande do Sul
O Commercial (Rio Grande)
Correio do Povo (Porto Alegre)
Correio do Sul (Porto Alegre)
O Criador Paulista (São Paulo)
Diário do Rio Grande
Echo da Fronteira (Livramento)
A Estância (Porto Alegre)
Gazeta de Porto Alegre

O *Mercantil* (Porto Alegre)
A *Pátria* (Montevideo)
O *Propagador da Indústria Rio-Grandense* (Rio Grande)
La Razón (Buenos Aires)
A *Reforma* (Porto Alegre)
Revista Agrícola da Fronteira (Livramento)
Revista Agrícola do Rio Grande do Sul (Pelotas)
Revista de la Asociación Rural del Uruguay (Montevideo)
Revista do Arquivo Público do Rio Grande do Sul
Revista do 1°. Centenario de Pelotas
South American Journal (London)
O *Sul Rural* (Porto Alegre)
Times (London)
Zero Hora (Porto Alegre)

Other Sources

Abel, Christopher, and Colin M. Lewis, eds. *Latin America, Economic Imperialism and the State: The Political Economy of the External Connection from Independence to the Present.* London: Athlone Press, 1985.

Abente, Diego. "Foreign Capital, Economic Elites and the State in Paraguay During the Liberal Republic (1870–1936)." *Journal of Latin American Studies* 21, no. 1 (Feb. 1989): 61–88.

Adelman, Jeremy. "Agricultural Credit in the Province of Buenos Aires, Argentina, 1890–1914." *Journal of Latin American Studies* 22, no. 1 (Feb. 1990): 69–87.

————. *Frontier Development: Land, Labour, and Capital on the Wheatlands of Argentina and Canada, 1890–1914.* Oxford: Clarendon Press, 1994.

Alden, Dauril. "Late Colonial Brazil, 1750–1808." In Bethell, ed., *CHLA*, 2: 601–60, 876–82.

————. *Royal Government in Colonial Brazil, with Special Reference to the Administration of the Marquis of Lavradio, Viceroy, 1769–1779.* Berkeley: University of California Press, 1968.

Alonso, José Antonio Fialho. "Análise do crescimento da região sul nas últimas décadas, 1959–90." In Alonso, Benetti, and Bandeira, *Crescimento econômico da região sul do Rio Grande do Sul*, pp. 51–93.

Alonso, José Antonio Fialho, Maria Domingues Benetti, and Pedro Silveira Bandeira. *Crescimento econômico da região sul do Rio Grande do Sul: causas e perspectivas.* Porto Alegre: Fundação de Economia e Estatística Siegfried Emanuel Heuser, 1994.

Antonacci, Maria Antonieta. *RS: as oposições & a revolução de 1923.* Porto Alegre: Mercado Aberto, 1981.

Araújo, Orestes. *Diccionario geográfico del Uruguay.* 2d ed. Montevideo: Tipo-Litografía Moderna, 1912.

Armstrong, Warwick, and T. G. McGee. *Theatres of Accumulation: Studies in Asian and Latin American Urbanization.* London: Methuen, 1985.

Assis Brasil, Joaquim Francisco de. *Cultura dos campos: noções geraes de agricultura e especiaes de alguns cultivos actualmente mais urgentes no Brasil.* 3d ed. Paris: Mounier, Jeanbin, 1910.

———. *Granja de Pedras Altas*. Buenos Aires: Talleres Heliográficos de Ortega y Radaelli, 1908.

———. *Guia do criador de carneiros*. Rio de Janeiro: Tip. da Cia. Nacional Editora, 1896.

Avé-Lallemant, Robert. *Reise durch Süd-Brasilien im Jahre 1858*. 2 vols. Leipzig: F. A. Brockhaus, 1859.

———. *Viagem pela província do Rio Grande do Sul*. Trans. Teodoro Cabral. Belo Horizonte: Editora Itatiaia; São Paulo: Editora da Universidade de São Paulo, 1980 [1859].

Azambuja, Graciano Alves de. *Anuário da Província do Rio Grande do Sul para o ano de 1888.* Porto Alegre: Krahe, 1887.

Azara, Félix de. "Memória rural do Rio da Prata." In Freitas, ed., *O capitalismo pastoril*, pp. 53–73. [The manuscript of this work, "Memoria rural del Rio de la Plata," dates from 1801.]

Baguet, Alexandre. "Rio Grande-do-Sul; tel qu'il était jadis et tel qu'il est actuellement." *Bulletin de la Société Royale de Géographie* (Antwerp) 18 (1893–94): 383–413.

———. *Rio Grande do Sul et le Paraguay*. Antwerp: Henri Ernest, 1874.

Baker, Alan R. H., ed. *Progress in Historical Geography*. Newton Abbot, Devon: David & Charles, 1972.

Bakos, Margaret Marchiori. *RS: escravismo & abolição*. Porto Alegre: Mercado Aberto, 1982.

Bandeira, Pedro Silveira. "As raízes históricos do declínio da região sul." In Alonso, Benetti, and Bandeira, *Crescimento econômico da região sul do Rio Grande do Sul,* pp. 7–48.

Barman, Roderick J. *Brazil: The Forging of a Nation, 1798–1852*. Stanford, Calif.: Stanford University Press, 1988.

Barrán, José P. *Apogeo y crisis del Uruguay pastoril y caudillesco, 1839–75.* Montevideo: Ediciones de la Banda Oriental, 1982.

Barrán, José P., and Benjamín Nahum. *Batlle, los estancieros y el Imperio Británico*. Vol. 1, *El Uruguay del novecientos;* vol. 2, *Un diálogo difícil, 1903–1910*. Montevideo: Ediciones de la Banda Oriental, 1979, 1981.

———. *HRUM*. 7 vols. Montevideo: Ediciones de la Banda Oriental, 1967–78.

———. "Uruguayan Rural History." *Hispanic American Historical Review* 64, no. 4 (Nov. 1984): 655–73.

Barreto, Abeillard. *Bibliografia sul-riograndense: a contribuição portuguesa e estrangeira para o conhecimento e a integração do Rio Grande do Sul.* 2 vols. Rio de Janeiro: Conselho Federal de Cultura, 1973–76.

Bauss, Rudy. "Rio Grande do Sul in the Portuguese Empire: The Formative Years, 1777–1808." *The Americas* 39, no. 4 (Apr. 1983): 519–35.

Beck, Mário de Lima. *Nova querência: chrónica das emigrações riograndenses para Matto Grosso*. Porto Alegre: Livraria Selbach, 1935.

Beek, Klaas Jan, and D. Luis Bramao. "Nature and Geography of South American Soils." In Fittkau et al., eds., *Biogeography and Ecology in South America,* 1: 82–110.

Bell, Stephen. "Aimé Bonpland and Merinomania in Southern South America." *The Americas* 51, no. 3 (Jan. 1995): 301–23.

―――. "Early Industrialization in the South Atlantic: Political Influences on the *Charqueadas* of Rio Grande do Sul Before 1860." *Journal of Historical Geography* 19, no. 4 (Oct. 1993): 399–411.

―――. "Foreign Investment and the Historical Geography of Brazil, 1850–1930: An Exploration of a Research Topic." M.A. research paper, Department of Geography, University of Toronto, 1980.

Benetti, Maria Domingues. "Agropecuária na região sul do Rio Grande do Sul, 1970–1990." In Alonso, Benetti, and Bandeira, *Crescimento econômico da região sul do Rio Grande do Sul*, pp. 95–211.

Bethell, Leslie, ed. *CHLA.* 10 vols. Cambridge: Cambridge University Press, 1984–.

Bishko, Charles Julian. "The Peninsular Background of Latin American Cattle Ranching." *Hispanic American Historical Review* 32, no. 4 (Nov. 1952): 491–515.

Blainey, Geoffrey. *The Tyranny of Distance: How Distance Shaped Australia's History.* Melbourne: Macmillan, 1968.

Blakemore, Harold, and Clifford T. Smith., eds. *Latin America: Geographical Perspectives.* London: Methuen, 1971.

Boerger, A., trans. G. M. Roseveare. "The Fundamental Factors of the Uruguayan Forage Problem." *Herbage Reviews* 7, no. 2 (June 1939): 70–79.

―――, trans. G. M. Roseveare. "Grassland Panorama of the La Plata Region." *Herbage Reviews* 6, no. 4 (Dec. 1938): 240–44.

Bonpland, Aimé. *Journal voyage de Sn. Borja a la cierra y a Porto Alegre: Diário viagem de São Borja à serra e a Porto Alegre.* Transcription of the original manuscripts, notes, and revision by Alicia Lourteig. Porto Alegre: Instituto de Biociências, Universidade Federal do Rio Grande do Sul; Paris: Centre National de la Recherche Scientifique, 1978.

Bourne, Richard. *Getulio Vargas of Brazil, 1883–1954: Sphinx of the Pampas.* London: C. Knight, 1974.

Boxer, C. R. *The Golden Age of Brazil, 1695–1750: Growing Pains of a Colonial Society.* Berkeley: University of California Press, 1962.

Brito, Severino de Sá. *Trabalhos e costumes dos gaúchos.* Porto Alegre: Companhia União de Seguros Gerais (ERUS), n.d. [Reprint of 1928 edition published in Porto Alegre by Livraria do Globo].

Broek, J. O. M. *The Santa Clara Valley, California: A Study in Landscape Changes.* Utrecht: Oosthoek's Uitg. Maatij., 1932.

Brookfield, Harold. *Interdependent Development.* London: Methuen, 1975.

Brossard, Paulo, ed. *Idéias políticas de Assis Brasil.* 3 vols. Brasília: Senado Federal; Rio de Janeiro: Fundação Casa de Rui Barbosa, 1989.

Brown, Gregory G. "The Impact of American Flour Imports on Brazilian Wheat Production: 1808–1822." *The Americas* 47, no. 3 (Jan. 1991): 315–36.

Brown, Jonathan C. "A Nineteenth-Century Argentine Cattle Empire." *Agricultural History* 52, no. 1 (Jan. 1978): 160–78.

————. *A Socioeconomic History of Argentina, 1776–1860*. Cambridge: Cambridge University Press, 1979.

Buarque de Holanda, Sérgio, gen. ed. *Historia geral da civilização brasileira*. 10 vols. São Paulo: DIFEL, 1960–75.

Burns, E. Bradford. *The Poverty of Progress: Latin America in the Nineteenth Century*. Berkeley: University of California Press, 1980.

Cabeda, Coralio Bragança Pardo. "José Antônio Martins, pioneiro esquecido do desenvolvimento da Campanha rio-grandense." *RIHGRGS* 130 (1994): 53–61.

Cardoso, Fernando Henrique. *Capitalismo e escravidão no Brasil meridional: o negro na sociedade escravocrata do Rio Grande do Sul*. 2d ed. Rio de Janeiro: Paz e Terra, 1977.

————. "Rio Grande do Sul e Santa Catarina." In Buarque de Holanda, ed., *História geral da civilização brasileira*, tome 2, vol. 2, *Dispersão e unidade*, pp. 473–505, 1964.

Carvalho, Mário Teixeira de. *Nobiliário sul-rio-grandense*. Porto Alegre: Globo, 1937.

Castilhos Goycochêa, Luiz Felippe de. *Eduardo de Araújo, Assis Brasil, Victor Russomano*. Porto Alegre: Tip. do Centro, 1941.

Castro, Juan José. *Treatise on the South American Railways and the Great International Lines Published Under the Auspices of the Ministry of Foment of the Oriental Republic of Uruguay and sent to the World's Exhibition at Chicago*. Montevideo: La Nación Steam Printing Office, 1893.

"Catálogo de manuscritos sobre o Rio Grande do Sul existentes na Biblioteca Nacional." *Anais da Biblioteca Nacional* (Rio de Janeiro) 99 (1979): 3–142.

Cesar, Guilhermino. *O conde de Piratini e a Estância da Música: a administração de um latifúndio rio-grandense em 1832*. Porto Alegre: Escola Superior de Teologia São Lourenço de Brindes and Instituto Estadual do Livro; Caxias do Sul: Universidade de Caxias do Sul, 1978.

————. *O contrabando no sul do Brasil*. Caxias do Sul: Universidade de Caxias do Sul; Porto Alegre: Escola Superior de Teologia São Lourenço de Brindes, 1978.

————. *História do Rio Grande do Sul: período colonial*. Porto Alegre: Globo, 1970.

————. "Ocupação e diferenciação do espaço." In Dacanal and Gonzaga, eds., *RS: economia & política*, pp. 7–28.

Chasteen, John Charles. "Background to Civil War: The Process of Land Tenure in Brazil's Southern Borderland, 1801–1893." *Hispanic American Historical Review* 71, no. 4 (Nov. 1991): 737–60.

————. *Heroes on Horseback: A Life and Times of the Last Gaucho Caudillos*. Albuquerque: University of New Mexico Press, 1995.

Chebataroff, J. "Regiones naturales del Uruguay y de Rio Grande del Sur." *Revista Uruguaya de Geografia* 2, no. 4 (1951): 5–40.

Cidade, Francisco de Paula. "Rio Grande do Sul: história explicada pela geografia." *Anais do III congresso sul-rio-grandense de história e geografia*, 2: 711–66. Porto Alegre: Globo, 1940.

Clark, Andrew H. "The Impact of Exotic Invasion on the Remaining New

World Mid-latitude Grasslands." In Thomas, ed., *Man's Role in Changing the Face of the Earth*, pp. 737–62.

Conde d'Eu (Prince Louis Gaston d'Orléans). *Viagem militar ao Rio Grande do Sul*. Belo Horizonte: Editora Itatiaia; São Paulo: Editora da Universidade de São Paulo, 1981 [reprint of 1936 edition].

II Congresso Nacional da Agricultura. *Conferencia feita em sessão de 20 de agosto pelo Dr. Assis Brasil*. Rio de Janeiro: Typ. do "Jornal do Commercio," 1908.

Cooper, Clayton Sedgwick. *The Brazilians and Their Country*. New York: Frederick A. Stokes, 1917.

Copstein, Raphael. "O trabalho estrangeiro no município do Rio Grande." *Boletim Gaúcho de Geografia* 4 (1975): 1–46.

Correa da Câmara, Antônio Manoel. "Ensaios statisticos da Provincia de S. Pedro do Rio Grande do Sul." *Revista Trimestral do Instituto Histórico e Geográfico da Provincia de São Pedro* 3, no. 2 (1863): 20–28.

Cortesão, Jaime. *Do Tratado de Madri à conquista dos Sete Povos (1750–1802): documentos*. Manuscritos da coleção de Angelis, no. 7. Rio de Janeiro: BN, Divisão de Publicações e Divulgação, 1969.

Cortés Conde, Roberto, and Stanley J. Stein, eds. *Latin America: A Guide to Economic History: 1830–1930*. Berkeley: University of California Press, 1977.

Costa, Emília Viotti da. "O escravo na grande lavoura." In Buarque de Holanda, ed., *História geral da civilização brasileira*, tome 2, vol. 3, *Reações e transações*, pp. 136–88, 4th ed., 1982.

Costa, Rogério Haesbaert da. *Latifúndio e identidade regional*. Porto Alegre: Mercado Aberto, 1988.

Coupland, Robert T., ed. (David W. Goodall, gen. ed.). *Ecosystems of the World*. Vol. 8A, *Natural Grasslands: Introduction and Western Hemisphere*. Amsterdam: Elsevier, 1992.

Couty, Louis. *Le maté et les conserves de viande: Rapport à son excellence Monsieur le ministre de l'agriculture et du commerce . . . sur sa mission dans les provinces du Paraná, Rio Grande et les états du Sud*. Rio de Janeiro: Typographia Nacional, 1880.

Crosby, Alfred W. *Ecological Imperialism: The Biological Expansion of Europe, 900–1900*. Cambridge: Cambridge University Press, 1986.

Crossley, J. Colin, and Robert Greenhill. "The River Plate Beef Trade." In Platt, ed., *Business Imperialism, 1840–1930*, pp. 284–334.

Curtin, Philip D. *The Rise and Fall of the Plantation Complex: Essays in Atlantic History*. Cambridge: Cambridge University Press, 1990.

Dacanal, José Hildebrando, and Sergius Gonzaga, eds. *RS: economia & política*. Porto Alegre: Mercado Aberto, 1979.

Daireaux, Godofredo. "Estancias in Argentina." In Argentine Republic, *Agricultural and Pastoral Census of the Nation*, 3: 1–51.

Darby, H. C. "Historical Geography." In Finberg, ed., *Approaches to History*, pp. 127–56.

———, ed. *An Historical Geography of England Before A.D. 1800*. Cambridge: Cambridge University Press, 1936.

Darwin, Charles. *The Voyage of the Beagle*. Garden City, N.Y.: Doubleday, 1962.

Dean, Warren. "The Green Wave of Coffee: Beginnings of Tropical Agricultural Research in Brazil (1885–1900)." *Hispanic American Historical Review* 69, no. 1 (Feb. 1989): 91–115.

———. "Latifundia and Land Policy in Nineteenth-Century Brazil." *Hispanic American Historical Review* 51, no. 4 (Nov. 1971): 606–25.

———. *Rio Claro: A Brazilian Plantation System, 1820–1920*. Stanford, Calif.: Stanford University Press, 1976.

Debret, Jean Baptiste. *Viagem pitoresca e histórica ao Brasil*. Belo Horizonte: Editora Itatiaia; São Paulo, Editora da Universidade de São Paulo, 1989. [Reprint edition of *Viagem pitoresca e historica ao Brasil: aquarelas e desenhos que não foram reproduzidos na edição de Firmin Didot—1834,* Paris: R. de Castro Maya Editor, 1954.]

Deffontaines, Pierre. "Contribution a une geographie pastorale de l'Amerique Latine: L'appropriation des troupeaux et des pacages." In L'Institut de géographie de l'université Laval, *Mélanges géographiques canadiens offerts à Raoul Blanchard*, pp. 479–91.

Downes, Earl Richard. "The Seeds of Influence: Brazil's "Essentially Agricultural" Old Republic and the United States, 1910–1930." Ph.D. dissertation, University of Texas at Austin, 1986.

Dreys, Nicolau. *Notícia descritiva da província do Rio Grande de São Pedro do Sul*. Reprinted with an introduction and notes by Augusto Meyer. Porto Alegre: Instituto Estadual do Livro, 1961 [Rio de Janeiro, 1839].

Dulles, John W. F. *Vargas of Brazil: A Political Biography*. Austin: University of Texas Press, 1967.

Duncan, K., and I. Rutledge, eds. *Land and Labour in Latin America: Essays on the Development of Agrarian Capitalism in the Nineteenth and Twentieth Centuries*. Cambridge: Cambridge University Press, 1977.

Duncan Baretta, Silvio Rogério. "Political Violence and Regime Change: A Study of the 1893 Civil War in Southern Brazil." Ph.D. dissertation, University of Pittsburgh, 1985.

Duncan Baretta, Silvio R., and John Markoff. "Civilization and Barbarism: Cattle Frontiers in Latin America." *Comparative Studies in Society and History* 20 (1978): 587–620.

Evans, Peter. *Dependent Development: The Alliance of Multinational,State and Local Capital in Brazil*. Princeton, N.J.: Princeton University Press, 1979.

Faria, Octavio Augusto de. *Diccionario geographico, historico e estatistico do estado do Rio Grande do Sul*. 2d. ed. Porto Alegre: Globo, 1914.

Feder, Ernest. *The Rape of the Peasantry: Latin America's Landholding System*. Garden City, N.Y.: Anchor Books, 1971.

Ferns, H. S. *Britain and Argentina in the Nineteenth Century*. Oxford: Clarendon Press, 1960.

———. "Britain's Informal Empire in Argentina, 1806–1914." *Past and Present* no. 4 (Nov. 1953): 60–75.

Finberg, H. P. R., ed. *Approaches to History: A Symposium*. London: Routledge & Kegan Paul, 1962.

Fittkau, E. J., J. Illies, H. Klinge, G. H. Schwabe, and H. Sioli, eds. *Biogeography and Ecology in South America*. 2 vols. The Hague: Junk, 1968–69.

Fontoura, João Neves da. *Memórias*. Vol. 1, *Borges de Medeiros e seu tempo*. Porto Alegre: Globo, 1958.

Forbes, Ian L. D. "German Informal Imperialism in South America Before 1914." *Economic History Review*, 2d ser., 31, no. 3 (1978): 384–98.

Fortes, Amyr Borges, and João Baptista Santiago Wagner. *História administrativa, judiciária e eclesiástica do Rio Grande do Sul*. Porto Alegre: Livraria do Globo, 1963.

Frances, May. *Beyond the Argentine: or, Letters from Brazil*. London: W. H. Allen, 1890.

Franco, Sérgio da Costa. "A Campanha." In Kremer, org., *Rio Grande do Sul*, pp. 65–74.

——. "Esquema sociológico da fronteira." *Província de São Pedro* 15 (1951): 46–51.

——. *Júlio de Castilhos e sua época*. Porto Alegre: Editôra Globo, 1967.

——. *Origens de Jaguarão, 1790–1833*. Caxias do Sul: Universidade de Caxias do Sul; Porto Alegre: Escola Superior de Teologia São Lourenço de Brindes and Instituto Estadual do Livro, 1980.

Freitas, Décio, ed. *O capitalismo pastoril*. Porto Alegre: Escola Superior de Teologia São Lourenço de Brindes; Caxias do Sul: Universidade de Caxias do Sul, 1980.

Freyre, Gilberto. *The Mansions and the Shanties: The Making of Modern Brazil*. New York: Alfred A. Knopf, 1963.

Friedmann, John. *Regional Development Policy: A Case Study of Venezuela*. Cambridge: MIT Press, 1966.

——. *Urbanization, Planning, and National Development*. Beverly Hills, Calif.: Sage Publications, 1973.

Fundação de Economia e Estatística. *De Província de São Pedro a Estado do Rio Grande do Sul: censos do RS, 1803–1950*. Porto Alegre: Fundação de Economia e Estatística, 1981.

——. *Guia de artigos sobre a história econômica do Rio Grande do Sul e temas correlatos*. 2 vols. Porto Alegre: Fundação de Economia e Estatística Siegfried Emanuel Heuser, 1990–91.

Fundação Instituto Brasileiro de Geografia e Estatística. *Levantamento de recursos naturais*. Vol. 33, *Folha SH.22 Porto Alegre e parte das Folhas SH.21 Uruguaiana e SI.22 Lagoa Mirim: geologia, geomorfologia, pedologia, vegetação, uso potencial da terra*. Rio de Janeiro: IBGE, 1986.

——. *Produção da pecuária municipal*. Tome 4, vol. 16, *Regiões Sul e Centro-Oeste*. Rio de Janeiro: IBGE, 1988.

Galloway, J. H. "Agricultural Reform and the Enlightenment in Late Colonial Brazil." *Agricultural History* 53, no. 4 (Oct. 1979): 763–79.

——. "Brazil." In Blakemore and Smith, eds., *Latin America*, pp. 335–99.

——. *The Sugar Cane Industry: An Historical Geography from its Origins to 1914*. Cambridge: Cambridge University Press, 1989.

Garavaglia, Juan Carlos, and Jorge D. Gelman. "Rural History of the Río de la

Plata, 1600–1850: Results of a Historiographical Renaissance." *Latin American Research Review* 30, no. 3 (1995): 75–105.

Gauld, Charles A. *The Last Titan: Percival Farquhar, American Entrepreneur in Latin America*. Stanford: California Institute of International Studies, 1972.

Giberti, Horacio C. E. *Historia económica de la ganadería argentina*. Rev. ed. Buenos Aires: Ediciones Solar, 1981.

Gibson, Herbert. "The Evolution of Live-stock Breeding in the Argentine." In Argentine Republic, *Agricultural and Pastoral Census*, 3: 53–106.

Gonçalves Chaves, Antônio José. *Memórias ecônomo-políticas sobre a administração pública do Brasil*. Porto Alegre: Companhia União de Seguros Gerais (ERUS), 1978 [Rio de Janeiro, 1822–23].

Gootenberg, Paul. "Beleaguered Liberals: The Failed First Generation of Free Traders in Peru." In Love and Jacobsen, eds., *Guiding the Invisible Hand*, pp. 63–97.

Graham, Gerald S. "The Ascendancy of the Sailing Ship, 1850–85." *Economic History Review*, 2d ser., 9, no. 1 (1956): 74–88.

Graham, Richard. "Brazil from the Middle of the Nineteenth Century to the Paraguayan War." In Bethell, ed., *CHLA*, 3: 747–94, 906–13.

———. *Britain and the Onset of Modernization in Brazil, 1850–1914*. Cambridge: Cambridge University Press, 1968.

———, ed. *The Idea of Race in Latin America, 1870–1940*. Austin: University of Texas Press, 1990.

Grees, H., and G. Kohlhepp, eds. *Tübinger Geographische Studien*. Vol. 102, *Ostmittel- und Osteuropa: Beiträge zur Landeskunde (Festschrift für Adolf Karger, Teil 1)*. Tübingen: Selbstverlag des Geographischen Instituts der Universität Tübingen, 1989.

Grigg, D. B. *The Agricultural Revolution in South Lincolnshire*. Cambridge: Cambridge University Press, 1966.

———. *The Agricultural Systems of the World: An Evolutionary Approach*. Cambridge: Cambridge University Press, 1974.

Halperín-Donghi, Tulio. "Argentina." In Cortés Conde and Stein, eds., *Latin America*, pp. 49–162.

———. *Politics, Economics and Society in Argentina in the Revolutionary Period*. Cambridge: Cambridge University Press, 1975.

Hecht, Susanna. "Environment, Development and Politics: Capital Accumulation and the Livestock Sector in Eastern Amazônia." *World Development* 13, no. 6 (1985): 663–84.

Hemming, John. *Red Gold: The Conquest of the Brazilian Indians, 1500–1760*. Cambridge, Mass.: Harvard University Press, 1978.

Hennessy, Alistair. *The Frontier in Latin American History*. London: Edward Arnold, 1978.

Hinchliff, Thomas. *South American Sketches*. London: Longman, Green, Longman, Roberts & Green, 1863.

Hirschman, Albert O. *Development Projects Observed*. Washington, D.C.: Brookings Institution, 1967.

———. *The Strategy of Economic Development*. New Haven, Conn.: Yale University Press, 1958.

Hirst, Monica. "Um guia para a pesquisa histórica no Rio de Janeiro: os documentos privados nos arquivos públicos." *Latin American Research Review* 14, no. 2 (1979): 150–71.

Hobsbawm, E. J. *Industry and Empire: An Economic History of Britain Since 1750.* London: Weidenfeld and Nicolson, 1968.

Hoffenberg, H. L. *Nineteenth-Century South America in Photographs.* New York: Dover, 1982.

Humphreys, R. A., ed. *British Consular Reports on the Trade and Politics of Latin America, 1824–1826.* London: Royal Historical Society, Camden Third Series, LXIII, 1940.

Hunt, Robert (assisted by F. W. Rudler). *Ure's Dictionary of Arts, Manufactures, and Mines: Containing a Clear Exposition of Their Principles and Practices.* 7th ed. 4 vols. London: Longmans, Green, 1878–79.

L'Institut de géographie de l'université Laval. *Mélanges géographiques canadiens offerts à Raoul Blanchard.* Québec: Presses universitaires Laval, 1959.

Intendência municipal de Jaguarão. *Apontamentos para uma monographia de Jaguarão.* Porto Alegre: Livraria do Globo, 1912.

Isabelle, Arsène. *Voyage a Buénos-Ayres et a Porto-Alègre, par la Banda-Oriental, les Missions d'Uruguay et la Province de Rio-Grande-do-Sul.* Le Havre: Imprimerie de J. Morlent, 1835.

Jacob, Raúl. "Las consecuencias sociales del alambramiento entre 1872 y 1880." In Mourat, ed., *Cinco perspectivas historicas del Uruguay moderno,* pp. 123–51.

———. *Las consequencias sociales del alambramiento, 1872–1880.* Montevideo: Ediciones de la Banda Oriental, 1969.

James, Preston E., and C. W. Minkel. *Latin America.* 5th ed. New York: John Wiley, 1986.

Jarvis, Lovell S. "Predicting the Diffusion of Improved Pastures in Uruguay." *American Journal of Agricultural Economics* 63, no. 3 (Aug. 1981): 495–502.

Jefferson, Mark. "Pictures from Southern Brazil." *Geographical Review* 16, no. 4 (Oct. 1926): 521–47.

Jones, Charles. "Commercial Banks and Mortgage Companies." In Platt, ed., *Business Imperialism, 1840–1930,* pp. 17–52.

Jones, Clarence F. "Agricultural Regions of South America," part 2. *Economic Geography* 4, no. 3 (July 1928): 159–86.

———. "The Evolution of Brazilian Commerce." *Economic Geography* 2, no. 4 (Oct. 1926): 550–74.

———. *South America.* New York: Henry Holt, 1930.

Jordan, Terry G. *North American Cattle-Ranching Frontiers: Origins, Diffusion, and Differentiation.* Albuquerque: University of New Mexico Press, 1993.

Joslin, David. *A Century of Banking in Latin America: To Commemorate the Centenary in 1962 of the Bank of London & South America.* London: Oxford University Press, 1963.

Karasch, Mary C. *Slave Life in Rio de Janeiro, 1808–1850.* Princeton, N.J.: Princeton University Press, 1987.

Kay, Cristóbal. *Latin American Theories of Development and Underdevelopment.* London: Routledge, 1989.

Kerst, Samuel Gottfried. "Die brasilische Provinz Rio Grande do Sul: Ein Beitrag zur Länderkunde." *Neues Magazin der neuesten Reisebeschreibungen* (Berlin) 47 (1832): 45–120, 289–332.

Kirby, John. "Uruguay and New Zealand: Paths to Progress." *Revista Geografica* 107 (Jan.-June 1988): 119–49.

Kliemann, Luiza Helena Schmitz. "A ferrovia gaúcha e as diretrizes de 'ordem e progresso,'—1905–1920." *Estudos Ibero-Americanos* (Porto Alegre) 3, no. 2 (Dec. 1977): 159–249.

Kohlhepp, Gerd. "Donauschwaben in Brasilien: Sozial- und wirtschaftsräumliche Entwicklungsprozesse der Heimatvertriebenensiedlung Entre Rios in Paraná." In Grees and Kohlhepp, eds. *Tübinger Geographische Studien,* pp. 353–86.

———. "Strukturwandlungen in der Landwirtschaft und Mobilität der ländlichen Bevölkerung in Nord-Paraná (Südbrasilien)." *Geographische Zeitschrift* 77, no. 1 (1989): 42–62.

Kremer, Alda Cardozo, org. *Rio Grande do Sul: terra e povo.* 2d ed. Porto Alegre: Editôra Globo, 1969.

Langer, Erick D. *Economic Change and Rural Resistance in Southern Bolivia, 1880–1930.* Stanford, Calif.: Stanford University Press, 1989.

Latham, Wilfrid. *The States of the River Plate: Their Industries and Commerce; Sheep-farming, Sheep-breeding, Cattle-feeding, and Meat-preserving; Employment of Capital; Land and Stock, and Their Values; Labour and its Remuneration.* London: Longmans Green, 1866.

Laytano, Dante de. "A colonização açoriana no Rio Grande do Sul." In Paula, org., *Colonização e migração,* pp. 391–421.

———. *Manual de fontes bibliográficas para o estudo da história geral do Rio Grande do Sul: levantamento crítico.* Porto Alegre: Universidade Federal do Rio Grande do Sul, 1979.

Leff, Nathaniel H. "Economic Development and Regional Inequality: Origins of the Brazilian Case." *Quarterly Journal of Economics* 86, no. 2 (May 1972): 243–62.

———. *Underdevelopment and Development in Brazil.* Vol. 1, *Economic Structures and Change, 1822–1947*; vol. 2, *Reassessing the Obstacles to Economic Development.* London: Allen & Unwin, 1982.

Leitman, Spencer Lewis. "The Black Ragamuffins: Racial Hypocrisy in Nineteenth Century Southern Brazil." *The Americas* 33, no. 3 (Jan. 1977): 504–18.

———. *Raízes sócio-econômicos da Guerra dos Farrapos: um capítulo da história do Brasil no seculo XIX.* Trans. Sarita Linhares Barsted. Rio de Janeiro: Edições Graal, 1979.

———. "Slave Cowboys in the Cattle Lands of Southern Brazil, 1800–1850." *Revista de História* (São Paulo) 51 (Jan.-Mar. 1975): 167–77.

———. "Socio-economic Roots of the Ragamuffin War: A Chapter in Early Brazilian History." Ph.D. dissertation, University of Texas at Austin, 1972.

Levine, R. M., ed. *Brazil: Field Research Guide in the Social Sciences.* New York: Columbia University, Institute of Latin American Studies, 1966.

Lewis, Colin M. *British Railways in Argentina, 1857–1914: A Case Study of Foreign Investment.* London: Athlone, 1983.

Lindman, C. A. M. *A vegetação no Rio Grande do Sul (Brasil austral).* Porto Alegre: Livraria Universal, 1906.

Lobb, C. Gary. "The Historical Geography of the Cattle Regions Along Brazil's Southern Frontier." Ph.D. dissertation, University of California at Berkeley, 1970.

———. "The *Sesmaria* in Rio Grande do Sul: A Successful Frontier Institution, 1737–1823." *Yearbook of the Association of Pacific Coast Geographers* 38 (1976): 49–63.

Lopes, Francisco Braziliense da Cunha, and João Luiz Nunes de Azevedo, orgs. *Carta geographica do Estado do Rio Grande do Sul* [Scale 1: 100,000]. Pelotas: Echenique, 1902.

Love, Joseph L. "History—Pôrto Alegre: Research Opportunities." In Levine, ed., *Brazil,* pp. 88–96.

———. *Rio Grande do Sul and Brazilian Regionalism, 1882–1930.* Stanford, Calif.: Stanford University Press, 1971.

———. *São Paulo in the Brazilian Federation, 1889–1937.* Stanford, Calif.: Stanford University Press, 1980.

Love, Joseph L., and Nils Jacobsen, eds. *Guiding the Invisible Hand: Economic Liberalism and the State in Latin American History.* New York: Praeger, 1988.

Luccock, John. *Notes on Rio de Janeiro, and the Southern Parts of Brazil: Taken During a Residence of Ten Years in that Country, from 1808 to 1818.* London: Samuel Leigh, 1820.

Luz, Nicia Villela. "Brasil." In Cortés Conde and Stein, eds., *Latin America,* pp. 163–272.

Lynch, John. *Argentine Dictator: Juan Manuel de Rosas, 1829–1852.* Oxford: Clarendon Press, 1981.

———. "The River Plate Republics from Independence to the Paraguayan War." In Bethell, ed., *CHLA,* 3: 615–76.

McColl, John. *The Republic of Uruguay, Monte Video, Geographical, Social and Political. To Which is Appended, Life in the River Plate. A Manual for Emigrants.* London: Effingham Wilson, 1862.

Macedo, Francisco Riopardense de. *Ingleses no Rio Grande do Sul.* Porto Alegre: A Nação, 1975.

Machado, Antônio Carlos. "A charqueada." *Província de São Pedro* 8 (1947): 121–36.

Maestri Filho, Mário José. *O escravo no Rio Grande do Sul: a charqueada e a génese do escravismo gaúcho.* Porto Alegre: Escola Superior de Teologia São Lourenço de Brindes; Caxias do Sul: Editora da Universidade de Caxias do Sul, 1984.

Magalhães, Manoel Antônio de. "Almanack da vila de Porto Alegre." In Freitas, ed., *O capitalismo pastoril,* pp. 74–102. First published as "Almanack da villa de Porto Alegre, com reflexões sobre o estado da capitania do Rio Grande do Sul," *Revista do Instituto Histórico e Geográfico Brasileiro* 34 (1867): 43–74 (the manuscript dates from 1808).

Malin, James C. *The Grassland of North America: Prolegomena to its History with Addenda and Postscript.* Gloucester, Mass.: Peter Smith, 1967 [1947].

Mansuy-Diniz Silva, Andrée. "Portugal and Brazil: Imperial Re-organization, 1750–1808." In Bethell, ed., *CHLA*, 1: 469–508.

Marchant, Anyda. *Viscount Mauá and the Empire of Brazil: A Biography of Irineu Evangelista de Sousa (1813–1889).* Berkeley: University of California Press, 1965.

Mariani, Alba A. "Los comienzos del proceso de mestización ganadera." In Mourat, ed. *Cinco perspectivas historicas del Uruguay moderno,* pp. 85–121.

Marques, Alvarino da Fontoura. *Episódios do ciclo do charque.* Porto Alegre: Editora e Distribuidora Gaúcha Ltda., 1987.

Marriner, Sheila. *Rathbones of Liverpool, 1845–73.* Liverpool: Liverpool University Press, 1961.

Martins, Ari. *Escritores do Rio Grande do Sul.* Porto Alegre: Ed. da Universidade Federal do Rio Grande do Sul and Instituto Estadual do Livro, 1978.

Mathias, Peter. *The First Industrial Nation: An Economic History of Britain, 1700–1914.* London: Methuen, 1969.

Mawe, John. *Travels in the Interior of Brazil; with Notices on its Climate, Agriculture, Commerce, Population, Mines, Manners and Customs: and a Particular Account of the Gold and Diamond Districts. Including a Voyage to the Rio de la Plata.* 2d. ed. London: Longman, Hurst, Rees, Orme and Brown, 1822.

Monte Domecq et Cie. *O Estado do Rio Grande do Sul.* Barcelona: Estabelecimento Grafico Thomas, 1916.

Mosk, Sanford A. "Latin America and the World Economy, 1850–1914." *Inter-American Economic Affairs* 2, no. 3 (winter 1948): 53–82.

Mourat, Oscar. "La inmigración y el crecimiento de la población del Uruguay, 1830–1930." In Mourat, ed. *Cinco perspectivas historicas del Uruguay moderno,* pp. 1–84.

——, ed. *Cinco perspectivas historicas del Uruguay moderno.* Montevideo: Fundación de Cultura Universitaria, 1969.

Mulhall, M. G., and E. T. Mulhall. *Handbook of the River Plate.* 6th ed. Buenos Aires: Standard Printing Office; London: Edward Stanford, 1892.

Müller, Jürg. *Brasilien.* Stuttgart: Ernst Klett Verlag, 1984.

Murray, Rev. J. H. *Travels in Uruguay, South America; Together with an Account of the Present State of Sheep-farming and Emigration to that Country.* London: Longmans, E. Stanford, 1871.

Myrdal, Gunnar. *Economic Theory and Underdeveloped Regions.* London: G. Duckworth, 1963.

Nicolau, Juan Carlos. "La industria saladeril en la confederacion argentina, 1835–1852." *Nuestra Historia* (Buenos Aires) 7 (Jan. 1970): 20–28.

Nunes, Zeno Cardoso, and Rui Cardoso Nunes. *Dicionário de regionalismos do Rio Grande do Sul.* Porto Alegre: Martins Livreiro, 1982.

Oberacker, Carlos H., Jr. "A colonização baseada no regime da pequena propriedade agrícola." In Buarque de Holanda, ed., *História geral da civilização brasileira,* tome 3, vol. 2, *Reações e transações,* pp. 220–44. 4th ed., 1982.

Oddone, Juan A. "The Formation of Modern Uruguay, c. 1870–1930." In Bethell, ed., *CHLA,* 5: 453–74.

Oliveira Belo, Luís Alves de. "Diário de uma excursão eleitoral feita pelo Dr. Luís Alves de Oliveira Belo ao deixar a Presidência da Província do Rio Grande do Sul." *RIHGRGS* 79 (1940): 5–45.

Oliven, Ruben George. "The Growth of Regional Culture in Brazil: An Analysis of the Resurrection of Gaúcho Identity in an Urbanized State." *Canadian Journal of Latin American and Caribbean Studies/Revue Canadienne des Études Latino-américaines et Caraïbes* 12, no. 23 (1987): 109–14.

Oribe Stemmer, Juan E. "Freight Rates in the Trade Between Europe and South America, 1840–1914." *Journal of Latin American Studies* 21, no. 1 (Feb. 1989): 23–59.

Paula, Euripedes Simões de, org. *Colonização e migração: anais do IV simpósio nacional dos professores universitários de história.* Coleção de Revista de História, 31. São Paulo, 1969.

Pebayle, Raymond. *Les Gauchos du Brésil: eleveurs et agriculteurs du Rio Grande do Sul.* Talence: Centre National de la Recherche Scientifique and Centre d'Etudes de Géographie Tropicale, 1977.

Pesavento, Sandra Jatahy. *República velha gaúcha: charqueadas, frigoríficos, criadores.* Porto Alegre: Ed. Movimento and Instituto Estadual do Livro, 1980.

———. "República velha gaúcha: estado autoritário e economia." In Dacanal and Gonzaga, eds., *RS: economia & política,* pp. 193–228.

Pfeifer, Gottfried. *Beiträge zur Kulturgeographie der Neuen Welt: Ausgewählte Arbeiten von Gottfried Pfeifer.* Berlin: Dietrich Reimer Verlag, 1981.

———. "Kontraste in Rio Grande do Sul: Campanha und Alto Uruguai." In Pfeifer, *Beiträge zur Kulturgeographie der Neuen Welt,* pp. 273–305 [first published in *Geographische Zeitschrift* 55, no. 3 (1967): 163–206].

Pimentel, Fortunato. *Aspectos gerais de Pelotas.* Porto Alegre: Tip. Gundlach, 1940.

———. *Joaquim Francisco de Assis Brasil, emérito agricultor.* Porto Alegre: Estab. Gráfico Santa Teresinha, 1950.

Pinheiro, José Feliciano Fernandes (Visconde de São Leopoldo). *Anais da Província de S. Pedro.* Rio de Janeiro: Imprensa Nacional, 1946. [First published in 2 vols. as *Anais da Capitania de São Pedro,* Rio de Janeiro: Imprensa Régia, 1819, and *Anais da Província de São Pedro,* Lisbon: Imprensa Nacional, 1822.]

Platt, D. C. M. *Latin America and British Trade, 1806–1914.* London: A. & C. Black, 1973.

———, ed. *Business Imperialism, 1840–1930: An Inquiry Based on British Experience in Latin America.* Oxford: Clarendon Press, 1977.

A Praça do Commercio da Cidade de Pelotas. *A praça do commercio da cidade de Pelotas ao Governo Imperial: representação.* Pelotas: Correio Mercantil, 1880.

Reber, Vera Blinn. *British Mercantile Houses in Buenos Aires, 1810–1880.* Cambridge, Mass.: Harvard University Press, 1979.

Reverbel, Carlos. *Um capitão da Guarda Nacional: vida e obra de J. Simões Lopes Neto.* Caxias do Sul: Universidade de Caxias do Sul; Porto Alegre: Martins Livreiro, 1981.

———. *Pedras Altas: a vida no campo segundo Assis Brasil.* Porto Alegre: L & PM Editores, 1984.

———, ed. *Diário de Cecília de Assis Brasil: período 1916–1928.* Porto Alegre: L & PM Editores, 1983.

Reyes Abadie, Washington, and Andrés Vázquez Romero. *Cronica general del Uruguay.* No. 54, *Los logros de la modernización,* and no. 61, *Población, comunicaciones y desarrollo urbano hacia 1900.* Montevideo: Ediciones de la Banda Oriental, n.d.

Ridings, Eugene. *Business Interest Groups in Nineteenth-Century Brazil.* Cambridge: Cambridge University Press, 1994.

Riet, Delfino M. *O cavalo crioulo—problema de defesa nacional.* Porto Alegre: Globo, 1918.

———. *Estância moderna.* Porto Alegre: Globo, 1926.

Ritvo, Harriet. *The Animal Estate: The English and Other Creatures in the Victorian Age.* Cambridge, Mass.: Harvard University Press, 1987.

Robinson, D. J. "Historical Geography in Latin America." In Baker, ed., *Progress in Historical Geography,* pp. 168–86.

Roche, Jean. *La colonisation allemande et le Rio Grande do Sul.* Travaux et mémoires, 3. Paris: Institut des Hautes Études de l'Amérique Latine, 1959.

———. *A colonização alemã e o Rio Grande do Sul.* 2 vols. Trans. Emery Ruas. Porto Alegre: Editôra Globo, 1969 [1959].

Rock, David. *Argentina, 1516–1987: From Spanish Colonization to Alfonsín.* Berkeley: University of California Press, 1987.

Rodrigues, J. H. *O Continente do Rio Grande.* Rio de Janeiro: Liv. São José Editora, 1954.

Roscio, Francisco João. "Compêndio noticioso." In Freitas, ed., *O capitalismo pastoril,* pp. 103–40. [First published as "Compêndio noticioso," *RIHRGS* 87 (1942): 29–56. The manuscript "Compendio noticiozo do Continente do Rio Grande de S. Pedro até o Destrito do Governo de Santa Caterina, extraido dos meus diários, observaçoens, e noticias, que alcancey nas Jornadas, que fiz ao ditto Continente nos annos de 1774, e 1775" dates from 1781.]

Roseveare, G. M. *The Grasslands of Latin America.* Bulletin no. 36. Aberystwyth: Imperial Bureau of Pastures and Field Crops, 1948.

Rouse, John E. *The Criollo: Spanish Cattle in the Americas.* Norman: University of Oklahoma Press, 1977.

Ryan, Shannon. *Fish out of Water: The Newfoundland Saltfish Trade, 1814–1914.* St. John's, Nfld.: Breakwater Books, 1986.

Safford, Frank. "Politics, Ideology and Society in Post-Independence Spanish America." In Bethell, ed., *CHLA,* 3: 347–421.

Saint-Hilaire, Auguste de. *Viagem ao Rio Grande do Sul, 1820–1821.* Trans. Leonam de Azeredo Penna. Belo Horizonte: Editora Itatiaia; São Paulo, Editora da Universidade de São Paulo, 1974 [1887].

———. *Voyage à Rio Grande do Sul, Brésil.* Orléans: H. Herluison, 1887.

Salvatore, Ricardo D. "Modes of Labor Control in Cattle-Ranching Economies: California, Southern Brazil, and Argentina, 1820–1860." *Journal of Economic History* 51, no. 2 (June 1991): 441–51.

Sanderson, Steven E. *The Politics of Trade in Latin American Development.* Stanford, Calif.: Stanford University Press, 1992.

Santos, Corcino Medeiros dos. *Economia e sociedade do Rio Grande do Sul: século XVIII.* São Paulo: Editora Nacional; Brasília: Instituto Nacional do Livro and Fundação Nacional Pró-Memória, 1984.

Sbarra, Noel H. *Historia del alambrado en la Argentina.* Buenos Aires: Raigal, 1955.

Scobie, James R. *Argentina: A City and a Nation.* 2d. ed. New York: Oxford University Press, 1971.

———. *Buenos Aires: Plaza to Suburb, 1870–1910.* New York: Oxford University Press, 1974.

———. Review of *Rio Grande do Sul*, by Joseph L. Love. In *Hispanic American Historical Review* 52, no. 4 (Nov. 1972): 683–84.

———. *Revolution on the Pampas: A Social History of Argentine Wheat, 1860–1910.* Austin: University of Texas Press, 1964.

Seckinger, Ron. *The Brazilian Monarchy and the South American Republics, 1822–1831: Diplomacy and State Building.* Baton Rouge: Louisiana State University Press, 1984.

Seckinger, Ron, and F. W. O. Morton. "Social Science Libraries in Greater Rio de Janeiro." *Latin American Research Review* 14, no. 3 (1979): 180–201.

Silva, Elmar Manique da. "Ligações externas da economia gaúcha (1736–1890)." In Dacanal and Gonzaga, eds., *RS: economia & política*, pp. 55–91.

Silva, Riograndino da Costa e. *Notas à margem da história do Rio Grande do Sul.* Porto Alegre: Editôra Globo, 1968.

Simões Lopes Neto, João. "Noticia sobre a fundação das xarqueadas." *Revista do 1°. Centenario de Pelotas* 2 (25 Nov. 1911): 10–12 and 3 (30 Dec. 1911): 44–46.

Skidmore, Thomas E. "Racial Ideas and Social Policy in Brazil, 1870–1940." In Richard Graham, ed., *The Idea of Race in Latin America*, pp. 7–36.

Slatta, Richard W. *Cowboys of the Americas.* New Haven, Conn.: Yale University Press, 1990.

———. "Gaúcho and Gaucho: Comparative Socio-economic and Demographic Change in Rio Grande do Sul and Buenos Aires Province, 1869–1920." *Estudos Ibero-Americanos* 6, no. 2 (Dec. 1980): 191–202.

———. *Gauchos and the Vanishing Frontier.* Lincoln: University of Nebraska Press, 1983.

Smith, Peter H. *Politics and Beef in Argentina: Patterns of Conflict and Change.* New York: Columbia University Press, 1969.

Solberg, Carl E. *The Prairies and the Pampas: Agrarian Policy in Canada and Argentina, 1880–1930.* Stanford, Calif.: Stanford University Press, 1987.

Soriano, Alberto, R. J. C. León, O. E. Sala, R. S. Lavado, V. A. Deregibus, M. A. Cauhépé, O. A. Scaglia, C. A. Velázquez, and J. H. Lemcoff. "Río de la Plata Grasslands." In Coupland, ed., *Ecosystems of the World*, pp. 367–407.

Sousa, José Antônio Soares de. "O Brasil e o Rio da Prata de 1828 à queda de Rosas." In Buarque de Holanda, ed., *História geral da civilização brasileira*, tome 2, vol. 3, *Reações e transações*, pp. 113–32. 4th ed., 1982.

Spalding, Walter. *A Revolução Farroupilha: história popular do grande decênio, seguida das efemérides principais de 1835–1845, fartamente documentadas*. 2d ed. São Paulo: Editora Nacional; Brasília: Instituto Nacional do Livro, 1980 [1939].

Stein, Stanley J. *Vassouras: A Brazilian Coffee County, 1850–1900*. Cambridge, Mass.: Harvard University Press, 1957.

Sternberg, Hilgard O'Reilly. "Man and Environmental Change in South America." In Fittkau et al., eds., *Biogeography and Ecology in South America*, 1: 413–45.

Street, John. *Artigas and the Emancipation of Uruguay*. Cambridge: Cambridge University Press, 1959.

Thomas, William L., Jr., ed. *Man's Role in Changing the Face of the Earth*. Chicago: University of Chicago Press, 1956.

Thompson, F. M. L. "The Second Agricultural Revolution, 1815–1880." *Economic History Review*, 2d ser., 21, no. 1 (Apr. 1968): 62–77.

Topik, Steven. *The Political Economy of the Brazilian State, 1889–1930*. Austin: University of Texas Press, 1987.

Uricoechea, Fernando. *The Patrimonial Foundations of the Brazilian Bureaucratic State*. Berkeley: University of California Press, 1980.

Vanger, Milton I. *José Batlle y Ordóñez of Uruguay: The Creator of His Times, 1902–1907*. Cambridge, Mass.: Harvard University Press, 1963.

Vereker, Hon. Henry Prendergast. *The British Shipmaster's Handbook to Rio Grande do Sul by the Honorable Henry Prendergast Vereker, LL.D., a Resident at Rio Grande do Sul*. London: Effingham Wilson, 1860.

Vianna, Lourival. *Imprensa gaúcha (1827–1852)*. Porto Alegre: Museu de Comunicações Social Hipólito José da Costa, 1977.

Vicuña Mackenna, Benjamín. *La Argentina en el año 1855*. Buenos Aires: Ed. de "la revista americana" de Buenos Aires, 1936.

Vidal de la Blache, Paul. "Les genres de vie dans la géographie humaine." *Annales de Géographie* 20, no. 111 (May 1911): 193–212 and no. 112 (July 1911): 289–304.

Villas-Bôas, Pedro. *Notas de bibliografia sul-rio-grandense: autores*. Porto Alegre, 1974.

Wade, Louise Carroll. *Chicago's Pride: The Stockyards, Packingtown, and Environs in the Nineteenth Century*. Urbana: University of Illinois Press, 1987.

Waibel, Leo. *Capítulos de geografia tropical e do Brasil*. 2d. ed. with annotations. Rio de Janeiro: Secretaria de Planejamento da Presidência da República, Fundação Instituto Brasileiro de Geografia e Estatística, Superintendência de Recursos Naturais e Meio Ambiente, 1979 [1958].

———, trans. Orlando Valverde. "Princípios da colonização européia no sul do Brasil." In Waibel, *Capítulos de geografia tropical e do Brasil*, pp. 225–77.

———, trans. Walter Alberto Egler. "As regiões pastoris do hemisfério sul." In Waibel, *Capítulos de geografia tropical e do Brasil*, pp. 37–62. [This article was originally published as "Die Viehzuchtsgebiete der südlichen Halbku-gels," *Geographische Zeitschrift* 28 (1922): 54–74.]

Walton, John R. "The Diffusion of the Improved Shorthorn Breed of Cattle in Britain During the Eighteenth and Nineteenth Centuries." *Transactions, Institute of British Geographers,* n.s., 9, no. 1 (1984): 22–36.

———. "Pedigree and the National Herd, *circa* 1750–1950." *Agricultural History Review* 34, part 2 (1986): 149–70.

Weimer, Günter. "Engenheiros alemães no Rio Grande do Sul, na década 1848–1858." *Estudos Ibero-Americanos* 5, no. 2 (Dec. 1979): 151–205.

Wendroth, Hermann Rudolf. *Album de aquarelas e desenhos, de cenas de viagem, vistas, tipos e costumes do Brasil, especialmente do Rio Grande do Sul,* c. 1852. Reproduced in the 1980's by the government of Rio Grande do Sul, n.p., n.d.; original unpublished manuscript in Arquivo da Casa Imperial do Brasil, Petrópolis, catalogue C, no. 218.

Whigham, Thomas L. "Cattle Raising in the Argentine Northeast: Corrientes, c. 1750–1870." *Journal of Latin American Studies* 20, no. 2 (Nov. 1988): 313–35.

———. *The Politics of River Trade: Tradition and Development in the Upper Plata, 1780–1870.* Albuquerque: University of New Mexico Press, 1991.

Wiederspahn, Henrique Oscar. *Bento Gonçalves e as guerras de Artigas.* Porto Alegre: Escola Superior de Teologia São Lourenço de Brindes; Caxias do Sul: Universidade de Caxias do Sul, 1979.

Wilcox, Robert. "Cattle and the Environment in the Pantanal of Mato Grosso, Brazil, 1870–1970." *Agricultural History* 66, no. 2 (spring 1992): 232–56.

———. "Paraguayans and the Making of the Brazilian Far West, 1870–1935." *The Americas* 49, no. 4 (Apr. 1993): 479–512.

Williams, John Hoyt. "The Archivo General de la Nación of Uruguay." *The Americas* 36, no. 2 (Oct. 1979): 257–68.

Winn, Peter. "British Informal Empire in Uruguay in the Nineteenth Century." *Past and Present* no. 73 (Nov. 1976): 100–126.

Winsberg, Morton D. "The Introduction and Diffusion of the Aberdeen Angus in Argentina." *Geography* 55, part 2 (Apr. 1970): 187–95.

Index

In this index an "f" after a number indicates a separate reference on the next page, and an "ff" indicates separate references on the next two pages. A continuous discussion over two or more pages is indicated by a span of page numbers, e.g., "57-59." *Passim* is used for a cluster of references in close but not consecutive sequence.

Library of Congress Cataloging-in-Publication Data

Bell, Stephen

    Campanha gaúcha : a Brazilian ranching system, 1850–1920 /
Stephen Bell.
      p. cm.
    Includes bibliographical references (p.   ) and index.
    ISBN 0-8047-3100-4 (alk. paper)
    1. Ranching—Brazil—Rio Grande do Sul—History—19th
century. 2. Ranching—Brazil—Rio Grande do Sul—History—
20th century. 3. Cattle—Technological innovations—Brazil—
Rio Grande do Sul—History—19th century. 4. Cattle—
Technological innovations—Brazil—Rio Grande do Sul—
History—20th century. 5. Rio Grande do Sul (Brazil)—
Economic conditions—19th century. 5. Rio Grande do Sul
(Brazil)—Economic conditions—20th century. I. Title.
SF196.B7B45   1998
636.2'.01'098165—dc21   98-19120
                                             CIP

This book is printed on acid-free, recycled paper.

Original printing 1998
Last figure below indicates year of this printing:
07 06 05 04 03 02 01 00 99 98